A BRILLIANT DARKNESS

A
BRILLIANT
DARKNESS

The Extraordinary Life and
Disappearance of Ettore
Majorana, the Troubled
Genius of the Nuclear Age

João Magueijo

A MEMBER OF THE PERSEUS BOOKS GROUP
New York

BASIC
BOOKS

Published by
Basic Books, A Member of the Perseus Books Group
387 Park Avenue South
New York, NY 10016

Books published by Basic Books are available at special discounts for bulk
purchases in the United States by corporations, institutions, and other
organizations. For more information, please contact the Special Markets
Department at the Perseus Books Group, 2300 Chestnut Street, Suite 200,
Philadelphia, PA 19103, or call (800) 255-1514, or e-mail
special.markets@perseusbooks.com.

Designed by Timm Bryson

Library of Congress Cataloging-in-Publication Data
Magueijo, João.
  A brilliant darkness : the extraordinary life and disappearance of Ettore
Majorana, the troubled genius of the nuclear age / João Magueijo.
      p. cm.
  Includes bibliographical references and index.
  ISBN 978-0-465-00903-9 (alk. paper)
  1. Majorana, Ettore. 2. Majorana, Ettore—Legends. 3. Nuclear physicists—
Italy—Biography. 4. Physics—Italy—History—20th century. 5. Neutrinos.
I. Title.
  QC774.M34M35 2009
  539.7092—dc22
  [B]
                              2009037678
10 9 8 7 6 5 4 3 2 1

CONTENTS

PROLOGUE
A Moment of Fatigue or Moral Discomfort

*Your highness: my son's malady has
been caused by noble studies.*

FROM A LETTER SENT TO MUSSOLINI
BY ETTORE MAJORANA'S MOTHER

In my early twenties, when I was still an apprentice scientist, I regularly worked as a scientific secretary at the Ettore Majorana Center in Sicily. This was always at the height of summer, when a brutal sun scorches the land, when you can't help but fall in love with the exuberant beauty of Sicily, a splendor full of superlatives, be it in the landscape, the people, or the food: *Beautiful* is never *"bello"* in Sicily, it's *"bellíssimo"*; *dangerous* is never *"pericoloso,"* it's *"pericolosíssimo."*

It was during one of these so-called "working trips," as a bunch of us sat outside demolishing yet another bottle of Corvo Rosso, that I first encountered Ettore. I refer to him by his first name now: He's been with me throughout my scientific career as a shadow I've never been able to shake off, perpetually reminding me of his story.

It was hardly surprising that Ettore's story came up during so many nighttime chats under starry skies: After all, we worked at a center named after him. For our efforts, we would be handsomely rewarded—in cash—by Il Centro Ettore Majorana. That bottle of wine, we wouldn't be paying for it: The center would settle the matter gracefully. Just as it took care of restaurant meals, nightclub fees, even the beach umbrellas at the nearby lido. If you're on the right side in Sicily, you don't offer payment: It might cause grave offense. And you certainly wouldn't want that.*

---

* There was a rumor that the banknotes used by the center were stained with blood. This is simply not true. They were extremely clean.

So Ettore was with us, at least in spirit, and I wasn't surprised to hear that he was not only a talented physicist but a Sicilian, born in the town of Catania. He'd been a nuclear physicist in Enrico Fermi's famous research group, the "Via Panisperna Boys," a whiz kid who could outdo even Fermi at the hardest calculations. In fact, Fermi compared Ettore to Isaac Newton and Galileo Galilei on a scale on which he considered himself second class—and on which he embarrassingly forgot to include Albert Einstein in the first rank.

However, this is not the end of the story. In fact, it's not even the beginning. Nowadays, Ettore is best known for something else: his final stunt. Without it, no one would want to know about his life. He'd be yet another victim of the myth that scientists are rational and emotionless—not real human beings. He'd be known for his theories about the neutrino and his contributions to nuclear physics and quantum mechanics, but nothing more.

Except for his stunt.

On the night of March 26, 1938, when Ettore was thirty-one years old, he boarded a ship in Palermo, Sicily, and was never seen again. He left behind a series of suicide notes and was known to have been depressed for at least five years; but what prevents the case from being closed is that his body was never recovered, and over the next few decades, he was allegedly sighted on numerous occasions. He also took with him the equivalent of $70,000, as well as his passport. It's hard to deny him human status after that.

Over the years, many theories have been proposed to explain what happened to him. Some claim he joined a monastery in Calabria; others that he may have run away to Argentina; still others say (or, rather, whisper) that he might have got into trouble with the Mafia. Conspiracy theorists consider him kidnapped by another political power eager for his nuclear knowledge. The lunatic fringe prefers him abducted by aliens, or perhaps still in flight in the extra dimensions. The fact is that no one really knows.

Naturally, on that Sicilian evening, fueled as we were by Corvo Rosso, and given that we were all budding physicists—a specie endowed with psychotic levels of imagination—we came up with even more fantastic theories. I recall someone suggesting that he'd bedded the Pope and been killed by a Vatican bodyguard.*

---

* Not as far-fetched as it might sound, I was assured.

But in all seriousness, Ettore's story still fascinated me the next morning—and has fascinated me ever since. The more I found out about him, the more relevant I found the issues raised by the story of his life. Over time, I became interested in more than just what happened to him; I became hooked on the question, what *drove* him to disappear? As my career progressed and I felt more acutely the many conflicts affecting scientists, the more I understood Ettore. Back in 1991, I was a raving idiot, too happy to be knowledgeable, too keen to be true. My feelings about Ettore grew with me as I matured and started to question the sanity of scientists myself. Until I began referring to him by his first name.

Ettore Majorana was born in Catania, on the eastern coast of Sicily, on August 5, 1906, at 8:15 pm. His early days were those of a child genius, a prodigy, a "gift from God." Having revealed a precocious ability with numbers, little Ettore was often paraded before visitors, doing cubic roots in his head, while the other kids were out playing marbles. He was never allowed to play and bore the corresponding solitude along with his brilliance.

In 1928, he abandoned conventional studies in civil engineering to become a theoretical physicist, at that time regarded as an unhinged pursuit. It was an impulsive act, never fully sanctioned by his family, but the poisonous bait was laid out for him by Enrico Fermi and his friends. Ettore at this time is described as Moorish looking, with intensely black hair and extremely bright eyes; disheveled and shy, constantly pondering, unable to chat—the caricature of a genius.

Now it just so happened that on Via Panisperna in Rome, there was a kindergarten for geniuses: a group of young, extremely bright physicists led by Fermi. They worked at an institute where they were given free rein: They held wagers as to who could solve differential equations the fastest, disputed audacious solutions to the riddle of the atomic nucleus, and dared each other to come up with the craziest theories of the universe. How they shouted . . . what a racket they made, even considering they were Italians. One day, one of them became so excited he threw a chair at another during an argument on the nature of solar fuel. Deep in their hearts, they were barely adults. In a sense, they *were* playing with marbles.

When Ettore first arrived at Via Panisperna to meet Fermi, the institute was struggling to solve what is now known as the universal Fermi potential, an essential tool for doing calculations in atomic physics. Fermi had managed to tabulate some regions of the potential, doing a giant number of sums essentially by hand (this was before computers), and although the results looked promising, they were still inconclusive. Fermi explained to Ettore why there were still such gaping holes in his table and why no one had been able to fill them. Ettore asked a few short questions, then left in the enigmatic style that would become his trademark.

The next day he returned and asked Fermi to show him the table again. Ettore then produced a piece of paper, did a few quick calculations, and congratulated Fermi on having made no mistakes. When Fermi looked surprised, Ettore went to the blackboard and wrote out a simple mathematical transformation converting the impenetrable problem into a well-known textbook equation. The picture of the full potential sprang into focus. Jaws duly dropped. Young Ettore was much given to theatricality.

The Via Panisperna Boys provided Ettore with an audience for his beloved magic tricks, which multiplied and became mythical, miraculous even, over time. Ettore's mathematical deftness created an aura of God-given talent, beyond human sweat; turned Ettore into the superhuman who simply "encountered" answers where others had to seek them with hard work—if they found them at all. And this is where his drama unfolded, until his dark side prevailed and he boarded that ship. As a scientist, I know the milieu well; I've seen the chairs fly during scientific arguments. And the thing that puzzles me most is not what happened to Ettore, but what made him so different from other physicists. Why did he alone feel so acutely the limitations of science?

"The moral of the scientist will always be unstable," said J. Robert Oppenheimer, the father of the atomic bomb, during a McCarthy-era hearing in which he was accused of being a risk to national security. Ettore was good friends with Werner Heisenberg, who later became head of the Nazi atomic project. Fermi (whose wife was Jewish) emigrated to the United States in 1939, where he led a line of research aiming toward the creation of the atomic bomb. Bruno Pontecorvo, another of the Via Panisperna Boys, assisted Fermi, but in 1950, smack in the middle of the cold war, he changed sides and absconded to the Soviet Union.

Ettore stands apart. His mental instability owes nothing to Hitler, McCarthy, or Stalin. There was always a major fault line that divided him from the Via Panisperna Boys. He particularly antagonized Fermi, who would be given the Nobel

Prize in 1938, the year of Ettore's disappearance. Ettore didn't care for prizes, was critical of academia, and always refused to be one of the boys. What made him play Il Grande Inquisitore (the Grand Inquisitor), as the Boys dubbed him?

Close friends and relations maintain that his instability derived from a dreadful incident implicating his uncles, dragging all to court and many to jail. At its epicenter lies a baby. A carbonized baby. Burnt by criminal hands, in an outburst of cruelty almost impossible to fathom.

There are so many holes in the story that we'll have to deal with it obliquely. But despite the labyrinth of question marks he left us, I've always seen Ettore as a role model. I firmly believe that he was endowed with a powerfully healthy soul, well kept and groomed despite trying times, in a world far less than perfect. The outcome was that he felt the need to vanish, leaving behind an impenetrable mystery. Shall we start with the scene of the crime?

On Friday, March 25, 1938, Ettore is at the end of his rope, filled with the kind of despair that invariably drives people to extremes. Holed up in his room at the Albergo Bologna in Naples, he pens two very strange notes, one to his boss and the other to his family. He's recently taken up a professorship at the University of Naples and is supposed to deliver a lecture on quantum mechanics the next morning. He's even spoken to one of his students about this lecture. But he never delivers it.

Instead, at 10:30 pm he boards a mail boat (a *piroscafo*) bound for the Sicilian capital of Palermo. The boat is due to arrive the next dawn. One of the notes he has written is addressed to the director of the Institute of Physics of Naples University, Professor Antonio Carrelli, who receives it on Saturday afternoon. It reads as follows:

Naples, 25 March 1938-XVI

Dear Carrelli,

I have made a decision that was by now inevitable. It doesn't contain a single speck of selfishness; but I do realize the inconvenience that my unanticipated disappearance may cause to the students and yourself. For this, too, I beg you to forgive me; but above all for having betrayed the trust, the sincere friendship and the sympathy you have so kindly offered me over the past few months. I beg you also to remember me to all those I've come to know and appreciate at your Institute, in particular Sciuti, of all I shall preserve the dearest memories at least until eleven o'clock this evening, and possibly beyond.

Ettore Majorana

His other note is addressed to his family and left on the desk in his room at Albergo Bologna. It's more laconic and reads:

(To my family)

Naples, 25 March 1938-XVI
    I've got a single wish: that you do not wear black for me. If you want to bow to custom, then bear some sign of mourning, but for no more than three days. After that remember me, if you can, in your hearts, and forgive me.

Ettore

Neither of these missives is a clear-cut suicide note. He uses the verb *scomparire*, which in Italian can mean either to die or to disappear. And then there's the handwriting: confident, secure, unwavering. Writer Leonardo Sciascia comments, "I've seen other suicide notes. In all of them there is more or less obvious alteration—even if only in the handwriting. Always. Some irregularity. Some confusion. Majorana's two notes reveal, on the contrary, composure, awareness, self-possession, a way of skirting ambiguity which, knowing him as I do now, cannot have been unintentional."

No one questions the peculiarity of these messages; but each person finds a different element particularly strange. Ettore's nephew Fabio, for example, tells me he attaches great importance to the odd grammar in the last phrase of the letter to the family. Ettore wrote very meticulously—mathematically even—and Italian grammar is considerably more structured than the English. And Ettore doesn't say, "Remember me and forgive me, if you can"; rather, he chooses to construct the phrase as "remember me, if you can, in your hearts, and forgive me." He is considering the possibility that the family might forget him. The emphasis is not on forgiveness, but on memory.

He never uses the verb "to die," and he never mentions death explicitly. He only talks of disappearing. As for wearing black, that can mean so many things. "For not more than three days" refers to the Sicilian custom. He's simply saying that the family should follow tradition if they want to keep up appearances, but no more than that.

I won't prejudice you any further with my own views: Reread these notes and form your own interpretations. You're guaranteed to see something new.

With or without intending suicide, Ettore boards the ferry in Naples that evening, but then the story takes an unexpected twist. Instead of "disappearing" as announced, Ettore disembarks in Palermo the next morning. He takes a room at the Albergo Sole and composes the following letter on the hotel's letterhead:

Palermo, 26 March 1938-XVI

Dear Carrelli,

I hope that my letter and telegram have reached you together. The sea has rejected me and tomorrow I'll return to the Hotel Bologna, perhaps traveling together with this same letter. I have, however, decided to give up teaching. Don't take me for an Ibsen heroine, because the case is quite different. I'm at your disposal for further details.

Ettore Majorana

The telegram referred to had in fact reached a very bewildered Professor Carrelli before Ettore's original letter, on Saturday morning. It states:

Don't be alarmed. A letter follows.
Majorana.

On Saturday evening, the piroscafo leaves Palermo on its return trip to Naples. There is some confusion about the details of this trip (even its exact date), but Ettore apparently bought a ticket, which was later recovered from the offices of Tirrenia, the shipping company. He's expected to arrive in Naples at 5:45 am on Sunday. There are two other people sharing the cabin in which he's supposed to be traveling, and later, one vouches that upon arrival in Naples, Ettore was still asleep in the cabin.

It looks as if the worst has passed—he has overcome a well-concealed low point in his depression. Ettore had never done anything of this sort before, even though he'd been very ill for a while. Now it seems he has given up on whatever drastic action he may have planned to solve his situation.

Instead, he's never seen again.

I want to make the case that the Nobel Prize for Physics should be given to Ettore Majorana. His science is easily in the same league as that of Einstein or Dirac. His disappearance may reek of defeat, but he's the real winner, the seer who could predict physics that is only now being tested. To this day I'm amazed that back in 1937, he succeeded in predicting the awkward behavior of a mysterious particle— the neutrino—in terms that are only now being probed.

It's almost a miracle that we ever discovered the existence of the neutrino, yet the universe is teeming with them. There are as many neutrinos out there as there are particles of light, and there are many, many more neutrinos than atoms or any other regular matter. Zooming around in all directions, journeying close to the speed of light, they're everywhere: swarming through everything, thousands of trillions hitting your body every second, coming from the skies, from below, from the horizon, from everywhere. If you've driven through the streets of Palermo, you may have noticed that a comparable number of scooters zip around your car in every conceivable direction, also flying past you at close to the speed of light, or so it seems. Driving in Palermo provides a perfect metaphor for neutrinos whooshing around us.

The analogy, however, ends there. Palermo's scooters occasionally embellish local cars with the end products of spectacular crashes; but neutrinos are way too shy for such exuberances. In fact, they're so modest and introverted that even though trillions hit your body every second, they pass through it as if you were a phantom. Hours pass before a single neutrino interacts with the atoms of your body. For neutrinos, the matter in the universe is perfectly transparent, diaphanous, immaterial; conversely, we fail to feel the effects of the colossal sea of neutrinos that envelops us.

Most of the neutrino hordes arriving on Earth come from the sun, yet the Earth is so transparent to them that at night we're showered by solar neutrinos *from below*, and the brightness of the sun in the "neutrino channel" is roughly the same night and day. The situation is so ridiculous that when we finally built the first neutrino telescope, we had to place it at the South Pole—to look at the northern sky! Using the whole of the Earth as a "lens" or filter, before miles of ice can act as the film in this peculiar neutrino camera.

We didn't detect a neutrino until 1956, using a very clever trick indeed. It had to be, since your regular neutrino can penetrate several light-years of ordinary matter as if it were traveling through a vacuum, without stopping to let itself be known.

How, then, could Ettore possibly already know about the neutrino back in the 1930s? The answer may very well have something to do with his disappearance. The neutrino may be shy, but it holds the key to humankind's nastiest weapon. Without the secret of the neutrino, nuclear weapons would not be possible. A massive burst of neutrinos spills out every time an atomic bomb goes off. These neutrinos are too soft and shy to cause any direct damage, yet they are instrumental to the conflagration.

The neutrino was first theorized by Wolfgang Pauli in 1930 to explain the mysterious properties of a type of nuclear radioactivity called beta decay. At Via Panisperna in the 1930s, no one doubted that the neutrino existed: Indeed, it was Fermi who christened it. The neutrino's name had to convey that it was electrically neutral, just like the neutron. "Neutron" phonetically becomes "neutrone" in Italian, and the suffix *one* is an augmentative. Thus, "neutrone" sounds like the "neutral big one." The suffix *ino* forms the diminutive*: Fermi thus dubbed the mysterious

---

* Those who enjoy life will recognize this pattern in the robust red wine Amarone and the light, strawberry-flavored Fragolino.

and very light extra particle the "neutrino"—the neutron's little brother. The neutrino formed a central part of the research at Via Panisperna.

And so it was that twenty-five years before the neutrino was even detected and proven to exist, Ettore discovered something extraordinary. Before "jumping ship" and disappearing from the face of the Earth, he left us a major piece of information about a particle that would play a leading role in the nuclear age. To use modern scientific parlance, he discovered that the neutrino could be Majorana. That's right, nothing less: Ettore has a neutrino named after him. Pending one small detail, an extant little mystery, he does indeed deserve the Nobel Prize.

I know that the Nobel Prize cannot be given posthumously: Statute 4 in its code makes that very clear. But is Ettore dead? We simply don't know. And nothing prevents the illustrious prize from being given in absentia. After all, Einstein collected his by proxy.

Sightings of Ettore began almost at once after his disappearance. As early as the end of March, start of April of 1938, Ettore reportedly approached a Jesuit priest, Father de Francesco, who peddled his trade at the church of Gesú Nuovo in Naples. Upon being shown a photograph of Ettore, the Father recognized him as the rather upset (*"agitadíssimo"*) young man who had inquired about the procedures for joining a monastery, in order to make an "experiment of religious life." Presented with the bureaucratic complexities of donating one's life to the Catholic God, however, the young man had left in a great hurry.

Then, in early April, Ettore's nurse, who'd helped him with a painful ulcer, claimed she saw him in Naples "on the street, between the Royal Palace and the Gallery"—hardly a secluded location. She recognized him by his face as well by his suit, indicating that he hadn't changed clothes since the day he disappeared.* If this story is true, it shows that Ettore made no great effort to vanish. Or else he was intelligent enough to know that often the best hiding spot is the most exposed location.

---

    * Not unusual in those days. A recent study on European laundry habits shows that, until recently, changing clothes was something done only on special occasions, like a purification process. This belief persists to this day among English academics.

On April 12, 1938, Ettore supposedly appeared at another monastery, San Pasquale a Portici, as reported by the local monks. This event was recorded in a letter from the *questore* (head of police) to the rector, showing that it must have formed part of the official inquiry into the disappearance that meanwhile had been opened. Unfortunately, this letter is one of the few official documents to have survived, and then only because the rector communicated it to the minister of education. All other documents pertaining to the police inquiry were destroyed, either as a matter of procedure or during World War II.

Encouraged by these early sightings, Ettore's brothers, Luciano and Salvatore, spent weeks scouring the countryside near Naples, trudging from monastery to monastery, showing Ettore's photo to the monks. But to no avail. From the religious they received replies such as, "But why do you search for him? Isn't the important thing that he's happy?" Still, they continued to seek. By then, however, sightings of Ettore had dried up.

Until some twenty years later. Suddenly, Ettore sightings became eerily frequent—a bit like sightings of Elvis Presley. He was seen in Argentina, at a convent in Basilicata, Italy, and in several less believable locations (Germany, western Sicily, even California). Ettore fanatics (Ettorologists) fall into three categories: Argentine theorists, suicide theorists, and monastery theorists. Were it not for Ettore's brilliant discovery, the Majorana neutrino, these conspiracy theorists would be the only force keeping his memory alive.

These days in Italy, everyone and his dog has written about Ettore Majorana. Ettore is even featured as the hero in several cartoon series. In one comic, he breaks into hyperspace and is visible only to cats (beasts allegedly well acquainted with extra dimensions). In another of his comic-book appearances, he discovers a quick and dirty way to make a fusion bomb.

Usually the H-bomb uses hydrogen, but Ettore's recipe calls for iron instead. This makes a devastating superweapon, one far more powerful than the H-bomb. He's acutely aware that everyone around him is way off track: Fermi; the rest of the Via Panisperna lot; the Russian, American, and German scientists. Ettore gets cold feet. He's scared to death of the Pandora's box he's unlocked: Think of such a lethal weapon in the hands of Mussolini or Hitler! But he talks to a friend, who convinces him that the Soviet Union—the "Land of the People"—is the only safe home for his invention. Ettore hesitates, but in the end goes along with his friend's plan. A Russian secret agent is dispatched to impersonate him on the boat bound for Naples. But Ettore never actually leaves Palermo, where another agent is scheduled to take him to a Russian fishing vessel stationed in international waters near Sicily.

But at the last minute, Ettore changes his mind. We see him sweating and panicking, shouting, "We're all doomed!" and writing to Professor Carrelli to cancel his earlier message.

Too late. The Russian spies and his "friend" now reveal themselves as evil. Ettore is led at gunpoint to a trawler at anchor nearby. The Russian secret agent acting as Ettore's double, meanwhile, duly convinces the few people who see him that Ettore boarded the boat to Naples, sending everyone off on the wrong trail.

Except that all of this is being closely monitored by the secret services of the "Land of the Free" (archenemy of the "Land of the People"). They have followed Ettore's activities and kept a close eye on the Russian agents. As soon as Ettore is safely aboard the trawler, they bring a submarine up to the surface. We see a commando platoon storming the defenseless craft, kidnapping for a second time the bewildered Ettore. They then torpedo the ship, sending all witnesses to the bottom of the ocean.

The kidnapped Ettore, however, refuses to participate in the American nuclear project, claiming that "they're all wrong; they're all doomed." He spends his time drawing diagrams no one understands and listening to long-wavelength radio transmissions that sound like static. They let him be, as long as he stays quietly in captivity. The bombs go off in Japan, the war is over, but nothing changes with Ettore. He continues to refuse to collaborate or beg for his freedom, indifferent to all around him.

Until July 24, 1947, when UFOs visit the Earth, targeting the military installation where Ettore is being kept. Earlier, guards had heard Ettore mumbling to the static on his radio, "The time has arrived, at last." The installation's defenses are activated, but the aliens discontinue the flow of time, freezing humans and animals in their tracks. The ground is littered with birds fallen from the sky. The flow of time has abandoned everyone—except Ettore.

Looking gloomier than ever, he wanders amid the frozen human statues. Full of emotion, Ettore approaches a silent, luminous flying saucer.

"I . . . I . . . I've been waiting for you."

He boards the flying saucer and leaves Earth for good.

Fifty years later, the CIA is using a special secret agent to communicate tele-pathically with alien worlds. In their midst, the telepathist sees Ettore. He's un-shaven and unkempt, but looks no older than when he disappeared. In a panic Ettore shouts to the telepathist: "The iron. They want to use iron, just like in my project. If they succeed in carrying out their plans, it will be the end!"

. . . If they succeed in carrying out their plans, it will be the END . . .

Very silly, indeed.

And yet, could this cartoon somehow have captured Ettore's real inner turmoil? Could this lowbrow take on his tragedy contain the perfect image of his suffering? It is now known that the Via Panisperna Boys accomplished the fission of uranium *without realizing it* as early as 1934. From there to the atomic bomb, all conceptual obstacles had been overcome. Did Ettore have an overdose of clarity? Or is even that too simplistic?

In real life, by the Wednesday after Ettore's disappearance, Professor Carrelli had given up on him, and sent the following letter to the rector of the University of Naples.

(Very confidential)
Naples, 30 March 1938-XVI

Magnificent Rector,
With great sorrow I communicate the following:

    Saturday, March 26 at eleven am, I received an urgent telegram from my colleague and friend Prof. Ettore Majorana, a Theoretical Physics Professor at this University, with a message composed in the following terms: "Don't be alarmed. A letter follows. Majorana." I found this missive incomprehensible, but made inquiries and discovered that he hadn't turned up for his lecture that morning. The telegram came from Palermo.

    With the two pm postal delivery I received a letter sent the day before, from Naples, where he expressed suicidal intentions. I understood then that the urgent telegram from Palermo, dated the following day, was intended to reassure me that nothing had happened. And indeed on Sunday morning I received an express letter from Palermo where he told me that his dark disposition was gone, and that he would soon return.

    Unfortunately, however, he didn't appear on Monday, either at the Institute or at the hotel where he had taken up lodgings. A bit alarmed by his absence, I sent news of all this to his family, resident in Rome. Yesterday morning his brother arrived, and together we went to see the questore of the City of Naples, to beg him to find out from the Palermo police whether Professor Majorana was still lodging at some hotel in that city.

    But since this morning I still had no news I decided to inform Your Magnificence of these occurrences, in the hope that my colleague has merely decided to take a little rest, following perhaps a moment of fatigue or moral discomfort, and that he will be rejoining us very soon to grace us with the great contribution of his activity and intelligence.

Antonio Carrelli

    "A moment of fatigue or moral discomfort" . . . this cannot have a happy ending, but you already know that. I warn you, however, that I'm going to be an optimistic influence throughout the narration of this pitiful tale.

For years after I first learned about Ettore, I wanted to write a book about his life. I collected documents and read nearly everything that had been written about him, but my research was slow and accretive, without much focus or purpose. I always deferred to an uncertain future the act of putting pen to paper. Then, one day, I read a newspaper clipping and realized that the mystery was about to come full circle. Ettore had just turned one hundred, and a major discovery had been made in the deep waters near Catania. The time had arrived for the final unveiling of the Majorana legacy.

The next day I caught a plane to Sicily. And so what if the mystery didn't crack after all? I'd always been in love with Sicily. The worst that could happen was a number of outstanding meals, severe sunstroke, perhaps death by beauty.

I . . . was waiting for you . . .

PART ONE

LIFE:
THE GRAND INQUISITOR

ONE

The Attic of 251 Via Etnea

❈ ❈ ❈ ❈ ❈ ❈ ❈ ❈

Where to start? For almost two decades I'd been baffled by Ettore's life, right from day one. So I began precisely at day one: Ettore's birth, his early days. That's usually where one finds the magic key to a person's later life, Freudian intellectual excrement aside. The question, then, was whether the "Rosebud" had already been thoroughly burnt and the ashes hygienically disposed of. I knew where he was born: at 251 Via Etnea in Catania. The top two floors of the imposing building are still occupied by Ettore's relatives. Before boarding that plane to Sicily, I found the contacts, sent e-mails, made phone calls. Honestly, I'd been fantasizing: Perhaps I would uncover a very aged Ettore hiding in the attic. In truth, I wasn't taking my chances very seriously: Why would a family carrying such a burden of past tragedy tolerate my intrusion? Still, it was worth a try.

To my pleased surprise, the doors of 251 Via Etnea open up for me. And it's Ettore's nephew Fabio, the son of his brother Luciano, who invites me in, leading me to the top floor for a chat. I can't quite believe my good luck—and I soon realize that things will likely be dramatic even if I don't discover Ettore in the attic.

At that moment, however, a terrible thing happens. All my previous sojourns in Sicily had been spent organizing conferences, ordering beer, chatting with locals . . . and I got by with working, if uncultured, Italian.* Suddenly, an appalling

---

* For a Portuguese native, Italian sounds like archaic speech. On my first trip to Sicily, I felt like a knight, talking to maids and other knights.

Signora Nunni Cirino Majorana, holding court at her house in Via Etnea.

realization dawns on me: My Italian is entirely limited to the present tense. I'm João the Pastless—so how will I discuss Ettore? Should I just ask, "Is Ettore in the attic?"

Undeterred, I converse with Fabio, using the most risible parody of his beautiful language. Remarkably, he doesn't laugh, even though at times it's obvious he's exercising considerable self-control. He takes me to the apartment where his mother lives, into a room full of red draperies, luxurious in a style that's at least a century old. I'm told that the furniture belonged to Ettore's maternal grandmother, who occupied the lower two floors of the house until she died. Sitting by the window is Signora Nunni Cirino Majorana, Ettore's sister-in-law, the last surviving member of his generation. When we enter, the signora is dozing, but she quickly wakes to greet us, evidently happy to have a visit. The signora is an old-fashioned, stylish lady—they don't make them like her anymore. She carries her eighty-two years with unequivocal dignity, a powerful twinkle in her eye, her wits still very much about her:

"I'm old and a bit of an invalid, but I have a fondness for life that has never been any greater. I love living!"

She pretends not to notice how exuberantly bad my Italian is:

"Signora is displayed like a young rose!"

"Would you like an iced tea?"

When she talks, there's a sharp clarity to her phrases, an intelligent humor that can't be far removed from Ettore's legendary subtle sarcasm. She's particularly keen to recall her long deceased husband, and it's via these recollections that I catch glimpses of what Ettore's family life must have been like.

"Signora wifed a brother of Ettore, Luciano?"

"I did, I did. Ah, that Luciano . . . he made me wait for over a year!"

"What age did signora owned at that time?"

"I was very young. And he was almost double my age. Everyone advised me against it, that he was way too old for me. But what did that matter to me? Why should I care? But for over a year I had to wait for him, not knowing if he'd marry me or not."

"Why he forced you to long, Signora?"

"That was the problem, more than his age! He said he couldn't do *that* to his mother; be so disrespectful, so tactless. I recall he even said that he couldn't be so *cruel.* . . ."

"But he was olded fifty years!"

She smiles (that beautifully mischievous smile!) the way she does when she thinks that silence is more eloquent than words. She makes a gesture around her lips signifying that they shall remain pursed.

"But in end he husbanded you."

"Yes, I waited for him."

"Was it worthed?"

That permanent shine in her eyes redoubles in intensity.

"Yes, it was well worth it! Every little bit." And then, as an afterthought, "His mother, Dorina, was a tough lady."

It's now Fabio's turn to talk:

"Grandma was good, had a very generous heart. But she was also *very* tough! No one argued with her: You just did what she said. When I was a child, she terrified me. And, naturally, she also terrified her own children."

The signora smiles, vaguely. Fabio lights up his pipe.

"She was a woman who knew she had to command for the family to move forward. She was full of initiative, of good sense, but with that she could also be overly protective."

He takes a drag. A beautiful fragrance fills the room. From outside comes the standard Italian racket of ambulance sirens, *motorini*, human chaos.

"It's true that her husband and children were useless with practical things. None of them, including her husband, could find his backside with both hands. But the fact is that she never accepted that her children would grow up. For her they were always children. She bought them pajamas; if they needed money, they'd ask her . . . and *this*, with sixty-year-old men!"

"I perused that Ettore lifed with mother until the year before his disappearance."

"That's right."

"She buyed him pajamas too?"

"Of course."

"How did mother react when she heard of his disappearance?"

There's an embarrassed silence. I realize I screwed up. But the signora finally says:

"Years later, how often I heard her say,"—her voice gains a threatening tone, mimicking what must have been Dorina's voice—"'When that Ettore returns, *he will hear me!*'"

Ettore's mother never believed that he died, it turns out. When she passed away at the age of ninety, many years after Ettore's disappearance, she left him his share in her will. Regardless of how much Ettore would "hear her" when he returned, the prodigal son would likely have been quickly forgiven. In Italy, Fabio tells me, you are officially declared dead if you've been missing for a couple of years. By the time his mother died, Ettore had been "officially dead" for well over twenty years.

I decide to go for the jugular, to plunge into what could be the "attic" where Ettore is hidden:

"Did Ettore have a strong impulse to evade away from his mother?"

This time, there's no embarrassed silence; the reply comes at once.

"Fortissimo!"

That was Fabio. The signora is nodding.

The next day, in the local paper, I read about an altercation that happened in Caltagirone, not far off. A very elderly lady confiscated her sixty-one-year-old son's keys, cut off his allowance, and dragged him to the police station because he stayed out late at night. At the station, the son protested that his mother didn't give him a large enough allowance and didn't know how to cook. "My son doesn't have any respect for me," the lachrymose lady was quoted as saying. "He doesn't tell me where he's going and returns home late. He hates my food and keeps on complaining. This can't go on."

The police helped them make up and they returned home together, with the son's keys and allowance restored.

Ettore was born in Catania, the second-largest Sicilian city after Palermo, with Mount Etna, one of the most active volcanoes in the world, looming in the background. Recently, the volcano has managed massive eruptions every three to four years, with minor affairs almost every year. (The cable car that goes to the top is one of the most expensive in the world: It has to be rebuilt every few years.) The eruption of 1669 reached Catania, almost twenty miles away, the lava filling its port and devastating the town. The city was resourcefully rebuilt in lava stone, its gorgeous baroque buildings now sporting a surreal dark gray hue. This would have lent it a heavy atmosphere were it not for the brutal Sicilian sun flooding its squares and avenues. Beyond the buildings, one can see Mount Etna smoking menacingly, but the pedestrians are far more worried about surviving the hysterical traffic.

Like everyone at the time, Ettore was born at home. A plaque, paid for by the Lions of Catania and the Etna, commemorates his birth at 251 Via Etnea. It reads, pompously:

In this house, on 5th August 1906, was born

ETTORE MAJORANA
Theoretical physicist

His timid and solitary genius scrutinized and illuminated the secrets of
the universe with the blaze of a meteor that too soon evaporated in
March 1938, leaving us the mystery of his thinking

The house at Via Etnea in Catania, where Ettore was born and raised.

Ettore's mother, Dorina, issued from a very rich family; indeed, that magnificent house on Via Etnea belonged to her side of the family—as did several other houses in Catania, as well as extensive vineyards at Passopisciaro, at the foot of Etna. Ettore's father, Fabio Massimo Majorana, is famous for setting up Catania's first telephone company, so the name Majorana used to sound to Catanians a bit like Bell

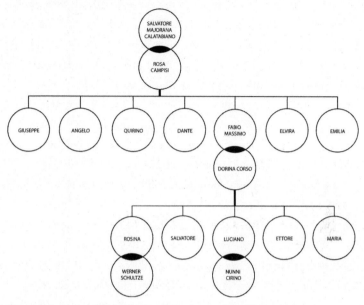

Figure 1.1: A very abridged genealogical tree of the Majorana family, covering three generations. The name of the out-of-family spouse is indicated wherever the branches are followed.

does in North America, synonymous with the telephone. As with all private enterprises, his venture was nationalized by Mussolini in the 1920s. Fabio Massimo received a pittance in compensation, but to sweeten the deal, he was appointed general inspector of communications, a cushy job in Rome. The family moved there in 1928.

If Dorina's ancestors bestowed vast wealth upon the family, prominence and political influence came from Ettore's paternal branch. The family jokes that their unusual name derives from Iulius Valerius Maiorianus, ordained emperor of the Western Roman Empire in AD 457; a jest, indeed, but one not totally divorced from reality, given the family's uncannily successful and precocious offspring (see Figure 1.1).

The founder of the modern Majorana dynasty was Ettore's grandfather, Salvatore, a self-made man who, while still very young, became an influential economist, lawyer, and writer, later becoming a member of Parliament; then minister of agriculture, industry, and commerce; and finally a senator. Out of nothing,* he propelled himself to the summit of academic and political achievement.

Contemporary sources rate Salvatore as a "left-wing" liberal, "the only person capable of unifying the various factions into which the left was split."† Other sources portray him as a man who succeeded out of sheer determination and through the power of his intelligence and "despite ancient class traditions and personal interests." Still others venture that he was "a man who antagonized the old powers, sometimes at a personal cost." The last reference, I believe, is to a stepson of his who was assassinated. A powerful baron was indicted for the crime and found guilty of ordering the murder, only to be promptly freed.‡

After the solid political foundations were laid by grandfather Salvatore, the family successes never ceased to multiply. Salvatore had seven children from his second marriage,§ and it's rather telling that three of them became rectors of the University of Catania (for the record: 1895 to 1898, 1911 to 1919, and 1944 to 1947). Following the path from academia to politics mapped out by their father, they also became members of Parliament, important public figures, and notable politicians.

One family member of particular note is Angelo, the second of Salvatore's sons, who was an example of absurd, if not obscene, precocity. Graduating with a degree in law at the age of sixteen, he was appointed assistant professor at the University of Catania at seventeen, publishing his first books the next year, to gain a full chair in constitutional law at twenty. He was appointed rector of the University of Catania at twenty-nine, a record to this day. Joining political life, he climbed to the powerful post of minister of finance in his thirties, to finally conquer the coveted position of minister of treasury, the second most powerful political post in the country.

For all his gifts, Angelo did something tremendously stupid at this point: He set out to solve Italy's financial and taxation problems (which still plague the coun-

---

* He married a rich widow.

† For reference, Italian political life—left and right wing—has always been a soap opera, one that fails to amuse the Italian people.

‡ The majority of the information on Ettore's family in this chapter originates from Erasmo Recami's book (see References at the end of the book).

§ His first marriage failed to produce any offspring.

try to this day). He soon fell violently ill, dying at the age of forty-four of terminal exhaustion. At 251 Via Etnea, a picture still hangs on the wall showing him at the apogee of his career—on his return to Catania after being appointed minister of treasury, on a carriage surrounded by a cheering crowd.

"I perused that Ettore's uncle, Angelo, acquired full professorness while aging at twenty," I say, the Italian language screaming in pain.

"It's remarkable, isn't it? He was still a minor; you officially became an adult at twenty-one in those days. When he took up his professorship, his father, Salvatore, had to sign for him because he couldn't legally do it. Imagine that! A full chair in constitutional law and he couldn't legally sign. . . ."

"Incredibly precocitous!"

"Yes, but you must realize they had to work extremely hard for it. Angelo and all his siblings," says the signora.

"I perused that they endured a verily tough regime."

"*Durissimo!* All up at six, they worked all day, following a strict plan of studies laid down by their father, interrupted only by the briefest of breaks for meals or to go to the bathroom, working late into the night before being packed off to bed. That's how the family successes were built. Those children didn't play."

"And that's how they educated their own offspring," Fabio continues. "Ettore and my father, before they were sent to school in Rome, studied long hours at home under the rigorous guidance of their father. They also didn't have much room to play."

We know what effect this had on Ettore—it turned him into a twisted prodigy—but it appears the story was quite different with his brother Luciano.

"My dad was not like the rest of the family, though. He never put pressure on us or forced us to study. He almost never talked of our past family glories. He used to say that as soon as you make children imitate their ancestors you kill their creativity. But in truth, I think he just didn't want to burden us."

The signora joins in pleasantly:

"My husband was really kind to the children. None of this nonsense Ettore and those other poor kids had to suffer. He was lighthearted, always playing with them, games, pranks, jokes. Always! It still warms my heart to think of their giggling filling the corridors of this house. . . ." She lowers her voice and her face changes into fake sternness, "Often at my expense."

"Why, signora?"

"Bah! . . . I could spend a whole afternoon recounting the horrors I had to suffer because of this lot!" She gestures toward an abstract location.

"For example, he'd stuff the kitchen tap with cardboard, then sit with the children until I turned up, and politely ask me for a glass of water, just so that they could all see me getting drenched."

She's enjoying her show of indignation (she's a great actress). Fabio and I are trying not to laugh.

"Another time, when I dozed off in a rocking chair, he made a circle with alcohol all around me and set it alight. Imagine my fright when I woke up! Until I heard them chuckling."

A few more stories along the same lines are reported, to Fabio's merriment. I get the idea that Luciano's pranks were legendary.

"But you liked it, signora."

"Me? What nightmares did I endure, because of this lot." She gestures toward the same undefined location. Her face then loses its fake rigidity and she joins us in our mirth. "Ah, my Luciano was so very kind to the children. . . ."

"Papa was a grown-up child," Fabio sums up.

Later I'm shown Ettore's and Luciano's school report cards. By this time they were attending a Jesuit school in Rome: They got marks for things like piety, civility, good behavior, urbanity . . . Ettore got 10 out of 10 in all; Luciano's card is a laughable parade of 5s and 6s.

"They had very diversified personalities," I comment.

"Is it too obvious?"

I realize I'm getting an image of the pressures Ettore must have endured *by the negative*: by learning about the way his brother reacted to those same pressures. In such an austere environment, you either become an anarchist or by the age of four you're doing cubic roots in your head and winning chess competitions. Ettore's early successes in chess, by the way, were reported—with awe—by a newspaper of the day.

Of Ettore's many uncles, I should highlight Zio Quirino Majorana (the third of Salvatore's sons), who was a physicist at Bologna University and corresponded with Ettore until the end. Today, Quirino Majorana is best known for never having believed in Einstein's theory of relativity, performing several complex experiments to disprove it. His attitude has been labeled "very Sicilian": He didn't care what the whole world thought; he believed that relativity was rubbish, and *he* was going to prove it. His fundamental honesty then surfaces in the fact that he never failed to report the negative results emerging from his experiments. But he kept on trying, stubbornly.*

Ettore's father, Fabio Massimo, is the fifth in Salvatore's remarkable litter, and he was definitely a man haunted by family expectations. But if his own political life was low-key, he was still a very successful businessman, despite being hopeless with money. By all accounts, Fabio Massimo was a disaster with practical things: His wife Dorina took the family reins. He looked after the children's education, but she collected money, took care of property sales and acquisitions, and kept track of the family's finances. If anyone needed any money in that house, they'd ask her.

Between such a mother and father Ettore grew up—predictably timid and eccentric, just like his siblings. One wonders how Luciano managed to be such an exception.

But I soon realize that if Luciano and Ettore had very different temperaments, as attested by their wildly different Jesuit-school report cards, they also had a lot in common. They shared a room until the year before Ettore's *scomparsa* (disappearance). On their school report cards, they both got 9s and 10s in the academic entries: study, diligence, benefit.

"Luciano was aeronautical engineer?"

"Yes, but he had to give it up. His mother was afraid of him flying and told him to quit. Which he did, unquestioningly."

---

* Leonardo Sciascia goes as far as to say that Ettore and his uncle "like all 'good' Sicilians" were averse to being part of *any* type of group, to establish teams or partnerships. He adds, "It's only the worst Sicilians who feel the need for 'groups,' for Mafia gangs."

"After that, he makes designs for telescopes, no?"

"Yes, he designed the solar telescope up on Etna, for example. And another two: one in Monte Mario, in Rome, and another at the Observatory of Gran Sasso in L'Aquila."

"He was a very curious man," the signora continues, "always interested in problems that might improve technology. For the sake of solving them, never for the sake of money!" she adds, as if she should have disapproved of her man's aloofness in financial matters; *should have*, but didn't.

"Father also prepared one of the first proposals for a bridge over the Messina Strait." I know what he means: The deep strait separating Sicily from the mainland is more than just a metaphorical chasm between cultures. It's been said that a bridge straddling it—a common promise made by lunatic politicians—would cost more than putting a man on Mars.

"Let me show you something," says Fabio, producing a letter from a huge folder. Recognizing it, the signora lets out some heartfelt laughter: "We should frame this treasure and put it on a wall!"

"In 1966, Papa sent Ferrari plans for a new type of car engine. It was built upon an idea he knew well from his days in the aeronautical industry. He'd designed engines for planes, different types of propellers and fuel feeding systems, and applied some of those ideas to cars. Why not? Back comes this brilliant document." He passes it on to me.

On Ferrari's letterhead is one of the most arrogant letters I've come across, and I've seen my fair share of scientific referee reports. Luciano is thanked for his endeavors, which have been well noted by Ferrari. However, he is informed, his contraption evidently could never be adapted to the complex engines built by that company; indeed one doubts that it could be of any practical significance whatsoever. Fabio waits for my inquisitive look before dropping the punch line:

"He'd sent them plans for a turbo engine."

The signora roars with laughter. Quite a few cars with turbo engines contribute to the din that comes from the street.

Later I find that Ettore's brother also worked on a prototype for an electrical train, with a similar response: He was ridiculed on the grounds that "such a train would require an electric wire running all the way along the track!" And in his attitude of laughing off the stupidity of the world, I find a common trait between Luciano and Ettore. They'd both get 10 out of 10 in "suffering poorly lesser intellects," in accepting rejection as part and parcel of originality. Ettore was a cheeky bastard: You can see it from the way he played with Fermi, right from day one, even on his induction day at Via Panisperna. I now see that it runs in the family.

We talk about originality and the scientific establishment, and I'm asked my opinion. I have strong views on the matter, so I manage to talk in Italian for ten long minutes. Listening to the tape, even I can't decipher what the hell I'm saying. I admire Fabio and the Signora's patience and politeness—their ability to keep a straight face.

I finally manage to steer the conversation away from my opinions.

"Why Uncle Quirino work in Bologna?"

"One day Papa Salvatore found out that he was dating a girl in Catania. I don't know how Quirino managed to slip out from his studies, but he did and even seduced a girl into the bargain! Over dinner, Salvatore simply ordered his son to pack, no reasons given, and sent him off to suitably monastic friends in Bologna. Dating girls wasn't permitted."

Gradually, I'm getting the picture of this unusual family. We're obviously talking about a clan with a strong binding energy, where the whole is more important than the parts, and where, during Ettore's generation, the head, the nerve center, was his mother, Dorina. I'm shown, not without pride, the cradle where Ettore slept as a baby. It's truly sweet—they used it for four generations: Ettore's father and uncles; Ettore and his four siblings; Luciano and the Signora's three children, including Fabio Jr. himself; and finally, as Fabio tells me, his own two children, now in their teens. That cradle symbolizes a bond that cuts through generations. Perhaps for this additional reason, setting fire to a cradle is such an aberration, such a stain. It would be a crime even before it's mentioned that a baby was sleeping in it at the time.

In one of the nearby bedrooms, Ettore was born. I was born at home, too. Being born at home creates a sense of attachment to a place that's absent in those who first see the light of day in a hospital. I can sense it at 251 Via Etnea: The house is haunted. Just as traces of urine from several generations pervade that cradle, the spirits of the past have entrenched themselves in the house.

The cradle where Ettore slept as a baby. It has served well four generations of disgruntled geniuses.

"How mother Dorina deaded?" I pidgin along.

I'm shown a photograph of Dorina's wake, her body spread on a bed as if she were asleep.

"Thus one dies," says the signora, looking at the photo.

"One day she got home," Fabio picks up, "and said she felt tired and didn't want any food. Salvatore and Maria, the children who still lived with her, at once panicked. They called the doctor and told him it was an emergency."

I laugh. She must have been hard as nuts. My grandfather was just like that.

"The doctor listened to her heart, and reported that it was nothing to worry about, she was just tired, that was all. But she looked him in the eye and said, 'Thus one dies.'"

"Thus one dies," echoes the signora.

"The doctor left. And two hours later, she was dead."

Signora is still nodding, as if by considering Dorina's words, she might be contemplating her own death.

"Fabio, you were the one to answer the phone, remember?"

There's an awkward silence; he doesn't take up the lead. I look more closely at him and see that there are tears in his eyes. The signora and I let the silence unroll.

"One night the phone rang, and I picked it up. I was a little kid. It was . . . I forget, it must have been Maria or Salvatore. They told me to go call Papa, that it was very urgent. I ran; Papa came to the phone ash-faced, as if he already knew the news. They hardly exchanged any words. He hung up, and I can never forget what he said at that moment."

His voice falters. The silence returns.

"He said, 'Now she knows where Ettore went.'"

For a while, none of us says anything. It's getting dark, but the heat coming from outside is still relentless. The signora and I let Fabio compose himself. He dries his eyes.

"You see, Papa never talked about Ettore to us. Maybe he felt it was too heavy going for children. But Ettore must have been permanently at the back of his mind. And of Dorina's mind. Something that silently affected them to the end of their days."

Looking at the photo of the dead Dorina, with all her children around her, I see the parallel with the cradle they all used through so many generations. Wakes are also times of family unity.

"Strange thing, this one of photographing dead people," says Fabio, looking at the picture.

Now she knows where Ettore went. . . . This woman who terrified three generations looks calm, relaxed, as if enjoying the most pleasant of dreams.

TWO

Nuclear Crisis

❧ ❧ ❧ ❧ ❧ ❧ ❧ ❧

Just to make things more theatrical, Ettore's science itself revolved around energy gone missing. Had he worked on time machines, conspiracy theorists couldn't have had more fuel. But it's a fact: For most of Ettore's life, no one knew where vast amounts of energy had gone. Energy was seemingly disappearing from the face of the Earth, unaccounted for, evaporated forever. Perhaps this might not sound so bad in Sicily, where "people have a knack for vanishing," as I was once told at a café in Catania. Skeptical, I picked up the local paper, *La Sicilia*. To the amusement of my interlocutor, right there on the front page was the *"storia triste"* of a certain Rita Cigno, *"la donna scomparsa,"* who'd vanished into thin air twelve years before. The family had enlisted the services of a clairvoyant who had just been jailed for fraud. I recalled Signora Nunni Cirino telling me how she'd been pestered by a similar professional of the occult.

Whether it's OK or not for people to vanish (not always to be found dead at the bottom of a pit), the very Sicilian Ettore was associated with an energy sink on a cosmological scale. But rather than add to this mystery, he helped solve it. Only to leave us with a larger conundrum, deep in the waters near Catania, as that other newspaper clipping—the one that prompted me to board a plane to Sicily and write this book—had hinted. Deep in those waters lies Ettore's Nobel Prize.

You already know what it's for: the neutron's sissy little brother, the neutrino.

✾

$E = mc^2$: Energy has mass, and mass is worth energy. That's what the most famous formula in the world tells us. The exchange rate is very large, the square of the speed of light, c^2, but conceptually matter and energy are equivalent: two sides of the same coin. Neither can be created or destroyed; they can only metamorphose, change in form and type, the overall amount of mass-energy always remaining the same. "In nature nothing is lost, nothing is gained, everything transforms," as in the adage of Antoine Lavoisier. When I launch a rocket into space, quite a lot of energy in the form of motion is gained, but it's not created out of nothing: It is born out of the chemical energy spent in burning fuel. When two particles collide and new ones are created, the *type* of mass-energy changes, but not its total amount. Conservation of mass-energy is one of the most sacrosanct principles of modern science, and it's as if Ettore's disappearance was the final twist in a long drama entailing energy destruction.

Yet that's precisely where Ettore's main discovery lies. It concerns "vanishing energy" and how it revealed a new particle, pathologically shy, almost impossible to detect: the neutrino. A perfect analogy between discoverer and discovered, observer and the observed, Ettore and his neutrino—an analogy spoiled only by the neutrino's lack of a domineering mother.

Instead, the neutrino had a domineering father: the often revered, always feared Wolfgang Pauli.

A brilliant Austrian physicist, "Pauli's mere presence in a city [was] enough to make all experiments go wrong," it was universally said. In the vicinity of theoretical physicists, his effect was similar: To him everything was *ganz falsch*, "utterly false," and he could be awfully nasty in his delivery of criticism. He was highly respected because his opinions were often, if not always, correct and also because he had equally high standards with regard to his own ideas. When he met the Dutch physicist Paul Ehrenfest, the latter felt compelled to remark, "I think I like your papers better than you." To which Pauli replied, "I think I like you better than your papers." The two became good friends.

A well-known anecdote describes Pauli, after his death, being granted an audience with God to discuss the secrets of the universe. After watching the old man scribble a profusion of formulae on the blackboard, Pauli explodes: *"Das is ganz falsch!"* Even God spoke out of His ass, according to Pauli.

But Pauli was also a complex man who drank way too much and endured long periods of extreme depression. His mother had committed suicide, and his father got remarried to a woman Pauli loathed, traumatizing him for the rest of his life. By 1930, Pauli had gone through several unhappy relationships, including a very short-lived marriage. (His wife dumped him for a chemist: "Such an average chemist," he commented.) In 1931, he had a severe nervous breakdown and became a long-term correspondent and friend of the psychoanalyst Carl Jung, who treated his "neurosis." Ettore was nicknamed Il Grande Inquisitore; Pauli was sometimes labeled the "conscience of physics." It seems that such roles come with a price.

In a historic letter dated December 4, 1930, Pauli wrote to the famous nuclear physicist Lise Meitner—"Dear Radioactive Ladies and Gentlemen"—and proceeded to apologize for his intended absence from an important meeting due to take place in Tübingen, Germany. He explained that "a ball which takes place in Zürich on the night of the 6th to 7th of December makes my presence here indispensable." The meeting in question was intended as a forum to discuss an extraordinarily grave problem in nuclear physics, something that had triggered a major crisis, casting doubt upon the respectability of the whole field of research. Pauli had sorted out his priorities correctly, favoring dance and flirtation over saving science from the perils of self-contradiction; but as a consolation in his letter to Meitner he offered a solution to the puzzle. "I dare not publish anything about this idea," he wrote, before branding his solution "a desperate way out."

In his letter, Pauli proposed a new particle—the neutrino, which he'd later call "that foolish child of the crisis of my life." A few months before, he'd divorced his wife after less than a year of married life; shortly afterwards, Jung had to record four hundred of his dreams before he was deemed "treated." Perhaps the ball he went to in Zürich wasn't wild enough; perhaps giving birth to an idiosyncratic new particle wasn't particularly fulfilling for him. As a memento of his frustrations, the particle has stayed with us.

With the massive benefit of hindsight—and simplifying a situation that was far, far more complex—I could shamelessly summarize nuclear physics in the early 1930s as follows: We're made of atoms, which have a nucleus with a positive

PROTON   NEUTRON   ELECTRON

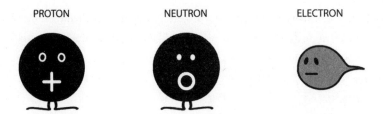

Figure 2.1: The three main characters in the universe. The proton and the electron have the same charge but with opposite signs (positive for the proton, negative for the electron.) The neutron is electrically neutral. The caricatures' mouths are meant as mnemonics.

ATOM

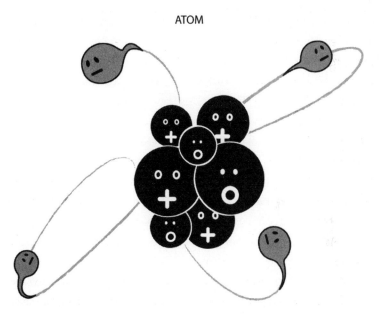

Figure 2.2: An atom as we understand it today (not to scale, and with a very large pinch of salt). The nucleus is made up of protons and neutrons. Electrons whizz around the nucleus at high speed.

charge surrounded by clouds of orbiting electrons with a negative charge (see Figures 2.1 and 2.2). The chemical properties of a substance result from the actions of the outer electrons and their ability to bring atoms together into molecules. Explaining the behavior of these electrons is the subject of *atomic* physics, but by the 1930s a new branch of physics had emerged concerned with the nucleus: *nuclear*

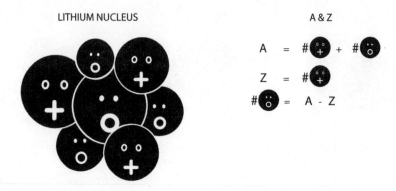

Figure 2.3: An illustration of the numbers A and Z labeling any element and isotope, where we chose lithium (A = 7 and Z = 3) as an example. The lithium nucleus is inhabited by three protons and four neutrons.

physics. This was to open up the world to a whole new "chemistry" no one had yet sampled.

The nucleus was known to contain two types of particles: protons, with a positive charge, and neutrons, electrically neutral (see Figure 2.1). Each type of nucleus, associated with each element in the periodic table, was characterized by two numbers: A and Z. A referred to the total number of protons and neutrons; Z to the total number of protons. The total number of neutrons was therefore A minus Z. For example, gold has A = 197 and Z = 79; that is, its nucleus has 79 protons and 118 neutrons. Lithium (used in many laptop batteries) has A = 7 and Z = 3 (see Figure 2.3).

In the early 1930s, we knew the world was made of the elements in the periodic table, but not exclusively: Radioactive isotopes were regularly being discovered. A given element in the periodic table has fixed A and Z values, but trace amounts of "twins" of these elements had been discovered in nature, with the same Z but a different A. For example, most oxygen atoms in nature have eight protons and eight neutrons, but 0.2 percent of oxygen atoms have eight protons and ten neutrons.

Such nonidentical twins were called isotopes. Because they had the same Z, and thus the same outer electrons responsible for chemistry, they had the same chemical reactions and other similar properties.* In an isotope, only the nucleus

---

* Note that since each atom must be electrically neutral, Z is also the number of electrons orbiting the nucleus. The proton and electron have the same electric charges, but with reverse signs.

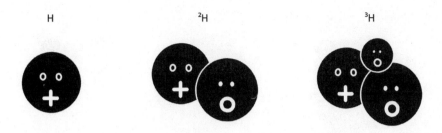

Figure 2.4: The three isotopes of hydrogen (Z = 1). Regular hydrogen (with A = 1) has only one proton in its nucleus. Deuterium (A = 2) has a proton and a neutron. Tritium (A = 3) has a proton and two neutrons. The nuclei of all these isotopes are orbited by a single electron (not represented) and so have the same chemical properties.

was different, because it contained a different number of neutrons. The overall mass per atom was therefore different, but other than that, chemistry was blind to the distinction between an element and its isotopes (see Figure 2.4).

These alternative versions of the elements existed in tiny amounts in nature, and it was difficult to "distill" or isolate them. But they had a striking property: Most of them naturally disintegrated into other elements or isotopes and in the process gave off especially energetic radiations. Because this phenomenon was first discovered in the element radium, it was called radioactivity. The process had astonishing features.

The energy associated with the radiations given off by radioactive disintegration was much, much higher than that in chemical processes: about a million times higher! Thus, those who studied nuclear physics must have realized at once that they had in their hands an enormous energy source. Explosives, at the time, resulted from particularly energetic chemical reactions. I wonder who first made this alarming connection.

Three types of radioactive reactions were known, associated with the three types of radiation that they gave off. They were labeled alpha, beta, and gamma. Alpha radiation was the least energetic and could easily be stopped with a sheet of paper. An alpha particle was made up of two neutrons and two protons, and it signaled the transmutation of a given element into another element with A reduced by four and Z reduced by two (see Figure 2.5).

Beta radiation, on the contrary, signaled a "sex change" in a neutron within the nucleus. Somehow a neutron became a proton, emitting very energetic radiation that could be stopped only by sheets of aluminum. Later, scientists understood that the stability of the nucleus is the result of balancing the number of neutrons

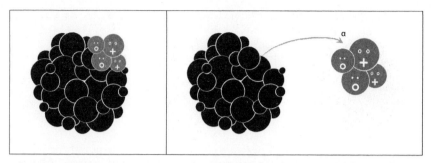

Figure 2.5: In alpha decay, a nucleus ejects an alpha particle (constituted by two protons and two neutrons), thereby losing two of its protons and two of its neutrons. After alpha decay, the nucleus transmutes into another with A reduced by four and Z reduced by two.

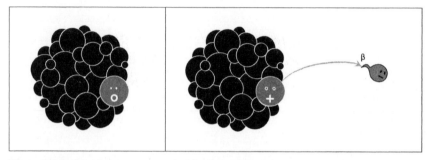

Figure 2.6: In beta decay, a neutron inside the nucleus converts into a proton, emitting a very energetic electron: the beta ray. Thus the daughter nucleus has the same A as the mother, but its Z is increased by one.

and protons (something Ettore helped explain, as we shall see). If an isotope had a surplus of neutrons with respect to the optimal number, it was liable to convert one of its neutrons into a proton, leading to a transmutation into an element with Z increased by one but with the same A value (see Figure 2.6). In the process, it emitted the very energetic beta radiation, soon identified with an electron moving close to the speed of light. Electrons emitted during beta radiation are not to be confused with those orbiting the nucleus, as they are emitted directly by the nucleus and are *much* more energetic.

By far the most powerful form of radiation was gamma radiation. This was an extraordinarily energetic particle of light, a photon (see Figure 2.7). No transmutation occurred as its result; rather, gamma radiation was emitted when the nucleus shook off surplus energy after one of the other types of radioactive disintegration had taken place. Gamma rays could be stopped only by very thick layers of lead

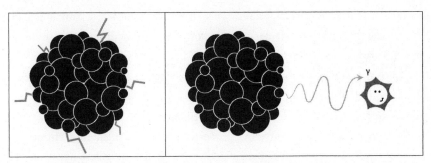

Figure 2.7: A gamma ray is a particle of very energetic light. It signals an energy offload by the nucleus, typically after undergoing one of the other types of decay. In gamma decay, the nucleus is simply letting off steam, and its A and Z remain unchanged.

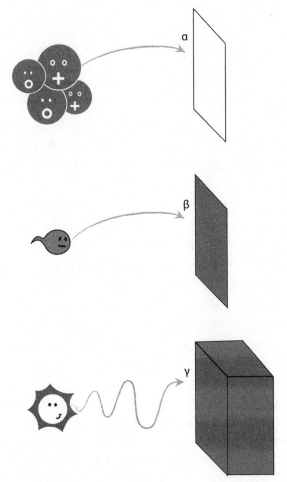

Figure 2.8: Alpha particles can easily be stopped by a sheet of paper. For beta rays, a sheet of aluminum might be a safer bet; whereas for gamma radiation, a thick layer of lead is the only protection against severe radiation damage.

and were by far the most biologically destructive of the radioactive radiations. Mutations, cancer, and tissue damage were among the horrors gamma rays could inflict (see Figure 2.8).

It was quickly understood that radioactivity was precisely the ancient dream of alchemists! It opened up the doors for a new type of "chemistry," in which elements transmuted into each other, for the number of protons and neutrons in nuclei did change during a radioactive process. By contrast, in chemical reactions, atoms merely changed their organization into molecules, or collections of atoms, but the nature of each individual atom remained unchanged. Turning sulfur into gold required nuclear physics. It was the "chemistry" of radioactivity that finally realized the original but failed motivation for chemistry—the philosopher's stone: the key to alchemical transmutation into gold.

What I've just described in a couple of pages originally took over thirty years to clarify. And although it's an oversimplification, with appropriate poetic license I can justifiably say that by the early 1930s, nuclear physics had pretty much fallen into place except for one tricky detail: the spectrum of beta decay. It was no laughing matter, and one must sympathize with the man who'd rather go to a ball than rack his brains about it. But, as usual, the problem was solved by the dancing grasshopper, while the ants labored at the lab. The matter that concerned all those ants, due to gather in Tübingen while Wolfgang Pauli licked his emotional wounds in Zürich, was simple: *Energy was disappearing.*

One can portray Ettore's 1938 eclipse as vanished energy, but that's only a metaphor. What was happening during beta decay was, by contrast, brutally concrete. A mother nucleus transmuted into a lighter daughter nucleus, and if you multiply their difference in mass by the square of the speed of light, according to the dependable $E = mc^2$, you obtain the amount of energy that *should* be carried away by each electron making up beta radiation. But the electron invariably had less energy. *Much less.**

---

* This is still true, even considering that the electron has to spend energy getting out of the atom and the material it is part of, and considering the recoil of the nucleus that fired it. Yes, people were not stupid then, and they did check all the details: We're talking about a million times more energy disappearing than could be accounted for by such details.

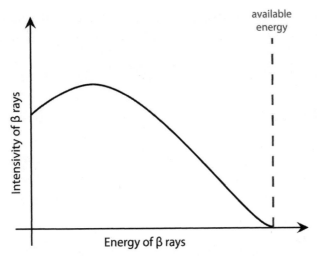

Figure 2.9: The spectrum of beta radiation. By considering the mass difference between mother and daughter nucleus, and using E = mc², one finds the "available" energy which should be carried by the beta ray. But it invariably carries less energy.

Where the hell was this energy going (see Figure 2.9)?

The energy, quite possibly, was going nowhere. It was just disappearing, and our principle of conservation of energy was wrong. Such was the opinion of Niels Bohr, who in a lecture given in May 1930 stated, point blank, "At the present stage of [nuclear] theory we may say that we have no argument, empirical or theoretical, for upholding the energy principle in the case of beta-ray disintegrations, and are even led to complications and difficulties in trying to do so." He was well aware of what he was insinuating—he added, "Of course a radical departure from this principle would imply strange consequences." However, he was prepared to take the leap and conclude, "[Still, the facts] may force us to renounce the very idea of energy balance."

I like Bohr's courage, but his was really a lazy explanation. After all, the same violations of energy conservation didn't plague the other radioactive processes, alpha and gamma emissions, and Bohr offered no explanation for this. This logical flaw excited Pauli's acerbic criticism; and thus in his December 1930 letter to Lise Meitner, Pauli suggested the presence of "a third man" to solve the mystery (see Figure 2.10). *Of course* energy was conserved—Bohr was a total donkey. It was just that an invisible third particle was taking it away. The neutrino was born.

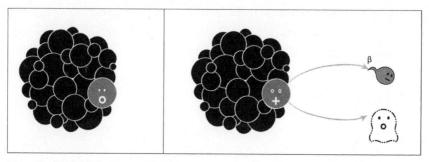

Figure 2.10: The full picture of beta decay according to Pauli. A ghost particle, the neutrino, carries away the missing energy.

What Pauli did is a common form of scientific witchcraft: He willed something into being by the power of logic and generalization. It's called making a prediction, and it is quite similar to what happens when a jealous wife finds that her husband is spending nights away from home. She postulates the existence of a lover even though she hasn't actually seen one. And, of course, in due time she's usually proved right.

For example, in the 1840s, French astronomer Urbain Le Verrier predicted the existence of the planet Neptune to explain strange perturbations in the orbit of Uranus. He claimed that this "Neptune" planet was too far away from us and therefore too faint to have been observed thus far, but that it must be there because its pull on Uranus was detected. He even directed astronomers as to where and when to look for the new planet. And when they pointed their telescopes to the appropriate spot they duly found the ghost planet, within a degree of Le Verrier's pronouncement. It was in this tradition that Pauli proposed the existence of the elusive neutrino. As Pauli put it, "I admit that my expedient may seem rather improbable from the first. . . . Nevertheless nothing ventured, nothing gained. . . . We should therefore be seriously discussing every path to salvation."

But after so much boldness, Pauli remained cautious about his creation. Admittedly he was suffering from a nervous breakdown, for unrelated reasons. But after his famous letter to Lise Meitner, excusing himself from the meeting and proposing the neutrino, the next time he aired his brainchild was on July 31, 1931, almost eight months later. Grossly overweight and in pain from a broken shoulder (having fallen down some stairs while drunk), Pauli finally owned up to his creation at a lecture in Pasadena, California. Even then he refused to write up his lecture; but a journalist attended it, and the *New York Times* reported on the neutrino hypothesis. For once, the overly keen science reporter was correctly ahead of the scientist.

We should not be too tough on Pauli, however. The luminaries of the time dubbed his idea "implausible," "crazy," or "simply wrong," and for a good reason. Some time later, Rudolph Peierls computed that so intangible was the neutrino that we'd need a few light-years of lead to stop it and make it interact with our detectors. That's a few million times the distance between the sun and Earth—*in lead*. So, the neutrino was not a ghost, but it was close.

Perhaps a bit like Ettore as a child.

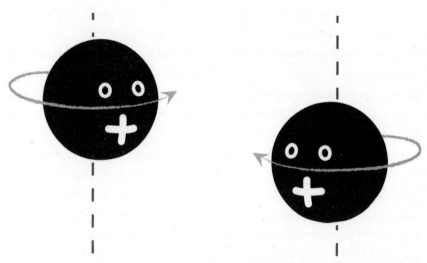

Figure 6.1: The spin of quantum particles, illustrated with a proton. Given a direction for the rotation axis (set, say, by an external magnetic field, here represented as a dashed line) the proton rotation "speed" is fixed, the only choice being whether it rotates one way or the other. The clockwise direction (left) is called spin up; the counterclockwise rotation (right) is referred to as spin down.

were discovered by means of "spin," a major player in Majorana's saga. Protons and electrons spin like tops, a property discovered in 1925 by two young Dutchmen, George Uhlenbeck and Samuel Goudsmit. But these spinning "tops" are unlike any other: Given a chosen direction for their rotation axis, their rotation "speed" is fixed, the only choice being whether they rotate one way or the other: "spin up" or "spin down" (see Figure 6.1). So important is spin to quantum mechanics that its central number, Planck's constant (which plays a role in quantum physics akin to that of the speed of light in relativity), is a measure of "spinning." In units of Planck's constant, the spin of the electron and proton is half: either plus or minus half, depending on whether it is spinning up or down.

Eerily, one can concoct superpositions of up and down, just like the ghost states of Schrödinger's cat. An electron can be 80 percent up and 20 percent down, and its future dynamics can be understood only if we assume that it evolves in this psychotic state. If you don't believe these oddities of "quantum spin" neither should you believe the results of your MR scan. Or trust your mobile phone, computer, TV set, automobile, etc. Dislike quantum mechanics all you want, but modern technology is predicated on it.

But where exactly is Lilliput? Just how small is "very small"? It can mean different things in different contexts, such as in the diverse subjects of atomic and nuclear physics. The two fields involve quite different creatures: Atomic physics deals with electrons whizzing around the nucleus; nuclear physics, with the dark secret of the nucleus itself. It wasn't until the 1930s that we first accessed the interior of the nucleus; the orbiting electrons, meanwhile, have been probed ever since the birth of chemistry.

The Boys first worked in atomic physics—it was their scientific nursery, where they honed their quantum-mechanics skills. We learn much about the electrons surrounding the nucleus from the light emitted when they transit between different quantum states: That's how quantum mechanics, with its package of oddities, was first revealed. Franco Rasetti was an expert in measuring this "spectral light," and Fermi was a leader in its mathematics. Segrè and Amaldi fawned upon these two, performing menial duties with enthusiasm and diligence. Ettore stood aside, his role, seemingly, to prove them all wrong.

Ettore's first known scientific intervention occurred in 1928, when he was only twenty-two years old. It was a communication to the Italian Physical Society, in which he corrected the Thomas-Fermi model, a way of dealing with multiple electrons in an atom by replacing them with "electron clouds." Fermi adamantly disagreed, which didn't stop him from publishing the same work six years later in collaboration with Amaldi, giving no credit to Ettore.

In his "atomic physics" period, Ettore published half a dozen papers, all single-author apart from one. Of these, his best-known work is the discovery of how "spectral light" shifts when atoms are bathed with an oscillating magnetic field. The result, the Majorana-Brossel effect, nowadays carries his name, and the mathematical technique used to solve the problem is called the Majorana sphere (still much in use; for example, in the beautiful formalism devised by Roger Penrose to describe spinning particles). Modern MR scans are an offshoot of Ettore's work.

But it was in nuclear physics that Ettore and, later, the Boys would leave their most indelible mark, beginning with Ettore's prediction of the neutron. Given their later role as experts on the topic, it's extraordinary that the Boys struggled to recognize the neutron's existence until as late as 1933. They weren't alone. At the time, almost without exception, scientists believed the nucleus to be made of protons and "nuclear" electrons. The idea was based on common sense. But common sense is seldom the path to truth.*

How were nuclei and atoms envisaged when Ettore disturbed the status quo? The hydrogen atom, for one, was easy to figure out. A hydrogen atom contains one nuclear proton orbited by one electron. The electron is much lighter than the proton— 1,836 times lighter—so that most of the atomic mass is in the nucleus: The mass of the hydrogen atom is roughly the mass of the proton. Because the proton and the electron have the same charge with opposite signs (positive for the proton, negative for the electron), the atom is electrically neutral. Nothing about the hydrogen atom was controversial; the issue was how to account for heavier nuclei.

Take nitrogen, for example. A nitrogen nucleus weighs roughly fourteen times as much as a proton, but it has a (positive) charge only seven times larger. The nucleus is orbited by seven electrons (rendering the atom neutral), but let's forget about those for the moment. Focusing on the nucleus, the question was how the nucleus could be fourteen times heavier than a proton but only have seven times its charge. The textbook answer was to employ fourteen protons and seven *intranuclear* electrons.† The charge comes out right because 14−7 = 7 (seven electrons neutralizing seven protons); and the mass also works out because the mass of the electrons is negligible, and there are fourteen protons in the nucleus (see Figure 6.2).

Believing the nucleus to be made of protons and electrons was therefore far from stupid, particularly considering that the fad of postulating thousands of unnecessary particles—as in string theory—was still well in the future. Combine this with the fact that during beta decay the nucleus emits one electron (suggesting that this electron already lived inside the nucleus), and you'd have to be an outright lunatic to come up with any other model for the nucleus. Which is exactly what Ettore did.

---

* The single voice of dissent came from Cambridge experimentalist Ernest Rutherford, who'd been hinting about the neutron since 1920. No one listened to him.

† That is, electrons that lived inside the nucleus, not to be confused with those orbiting the nucleus on the outer edges of the atom.

THE NITROGEN NUCLEUS THEN AND NOW

Figure 6.2: The nitrogen nucleus as it was understood before and after the neutron entered the picture. Charge seven and mass fourteen can be explained either with fourteen protons and seven nuclear electrons, or with seven protons and seven neutrons.

Because in reality there *was* a need for the neutron, as should have been obvious since the 1920s. Much as the Boys resisted it, of all people they should have known better. No one more so than Franco Rasetti, who was sent abroad by Fermi in 1929 "to season," and found himself at Caltech studying the properties of the nitrogen nucleus, just as in the example I chose. The Cardinal Vicar's project was to measure the spin of this nucleus.

When you combine spins in groups, they add or subtract according to well-defined rules. Two spin half particles can only add up to a spin one (when both spins align in the same direction: $\frac{1}{2} + \frac{1}{2}$) or spin zero (when the spins point in opposite directions: $\frac{1}{2}-\frac{1}{2}$). Quantum mechanics tells you the *probabilities* of one pairing over the other. And similar addition laws rule larger conglomerates of quantum particles. *Rasetti found that the nitrogen nucleus has spin one.*

This utterly contradicted a nucleus made of fourteen protons and seven electrons, because such a nucleus would have an odd number (twenty-one) of particles. No matter how you pair and combine an odd number of spin half particles, you can only produce a half-integer collective spin: $\frac{1}{2}$, $\frac{3}{2}$, $\frac{5}{2}$, etc. Yet the total spin of the nitrogen nucleus was one. A puzzling result—ignored by everyone and quickly forgotten.

Except by Ettore. He must have been the first to understand that if the nitrogen nucleus contained protons and a new kind of spin half particle—the neutron—then Rasetti's observation could be explained. In contrast to Amelio's film, Ettore did talk openly to the Boys about a "neutral proton": a particle with a mass similar to the proton, no electric charge, and spin half. He noticed that given its existence, the nitrogen nucleus could be made up of seven protons and seven neutrons, accounting for mass fourteen and charge seven. And then there would be an even number (fourteen) of spin half particles in the nucleus, so that their spins could add up to an integer number. Such as one, the observed value.

The clues were all there—at the Panisperna labs! And Niels Bohr, the father of quantum mechanics, had already remonstrated about "the remarkable passivity of intranuclear electrons," which somehow lost even their spin as soon as they were inside the nucleus. Yet these clues went unheeded, and Ettore did in fact "see beyond," as portrayed in Amelio's film, when he predicted the neutron's existence. But there was more to it than the nitrogen spin experiments: There was something else of crucial importance that may have scared him.

Rasetti's discovery may have been important, but there was at the time another much more blatant conundrum in nuclear physics: No one understood why a nucleus made of protons and electrons, governed only by electricity, didn't simply blow apart. The energy budget of a nucleus governed by electric attractions and repulsions favored its quick dismantling. Scientists envisioned the nucleus as a small molecule, which is indeed stable due to the collective electric attractions and repulsions of its components. But the nucleus is 100,000 times smaller than any molecule, and for something that small, electric repulsion wins in the overall budget. It just doesn't scale, as a simple calculation (surely performed by Ettore) quickly shows. In order to explain why the nucleus was bound together, a new type of force, stronger than electricity, was needed, as well as a new, electrically neutral particle.

We find this prescient remark in Ettore's notebooks when he was in his early twenties. He wasn't being stubborn: He'd tried to set up a nuclear model using the conventional picture and had come up with nonsense. He wasn't proposing a new particle and a new force gratuitously—he was seriously attempting to explain the mysterious stability of the nucleus. And he'd found that with his new particle and force, not only was the nucleus stable, but its binding energy was millions of times larger than that binding electrons in atoms. Did he gasp in horror upon recognizing the enormous strength of nuclear forces? We don't know. No one listened to him on the few occasions he mentioned neutrons and strong forces to the Boys.

The picture changed dramatically when the Joliot-Curie husband and wife partnership accidentally produced the first free neutrons. They had no idea what they'd found but, as good experimentalists, gave an accurate account of what they'd seen. This new radiation could easily approach or even penetrate nuclei and produced such gigantic knocks on these nuclei that its particles had to be at least as massive as the proton. Everyone, including the Joliot-Curies, was baffled. What the hell was going on?

It's telling that Ettore, upon hearing the news, chose to play the Inquisitor: "They haven't understood *anything!* It's probably the result of recoil protons produced from heavy neutral particles," Ettore said, according to Amaldi. The Joliot-Curie particle was like a proton but had to be neutral: If it were positively charged, it would have to overcome an enormous repulsion barrier to penetrate the (positively charged) nucleus. For someone like Ettore who had worked with neutrons for several years, the Joliot-Curie results were therefore hardly surprising.

Meanwhile, Fermi and the Boys were utterly in the dark. Even Werner Heisenberg, who officially proposed the first theory of nuclear stability, couldn't quite believe in the new particle. To illustrate the generalized mental block, I note that the great Werner understood the "neutron" as a "little atom" made of one proton and one electron. In effect, he simply repackaged the old nuclear model of protons and electrons, combining some of the electrons and protons into couples. Rasetti's experiment clearly hadn't sunk in. The neutron has spin half, whereas such couples would have spin one or zero. No wonder Ettore felt impatient.

Not only was Ettore sure of the neutron's existence, he had also finished a theory of nuclear stability (and isotope instability) based on protons and neutrons, bound by a new type of force he dubbed "force of exchange." This was the precursor of the strong force and later quantum chromodynamics (or QCD), and it was a most remarkable breakthrough. Ettore didn't burn it, but he never published it, vehemently refusing to do so to everyone's puzzlement. Although Fermi had yet to be convinced by the argument, he realized that Ettore's theory might be right and what a coup it would be for Via Panisperna. He insisted that Ettore publish it—to no avail.

So upset was Fermi that he even suggested presenting the theory at an important congress in Paris, giving due credit to Ettore. Ettore agreed, but on condition that the theory be attributed to a certain professor of electronics in Rome, widely regarded as a moron, who was bound to attend the conference. Fermi didn't have the guts for the prank. And so Ettore's idea remained in the drawer.

A few months later, Dmitri Ivanenko in Russia recognized the new particle, and Heisenberg in Germany proposed much the same theory as Ettore. When Fermi expressed his disappointment, Ettore only laughed. He found the entire situation hilarious. And when Fermi suggested publishing at least a follow-up, Ettore—having refused to publish before Heisenberg—said, "Now Heisenberg has already done everything." And he chortled away while the other Boys commiserated.

How can we ever understand Ettore's bizarre behavior? Was he disturbed by the large nuclear binding energies he'd found; did he foresee their deadly potential? Or are there simpler reasons? For someone as driven by competition as Fermi, Ettore must have been *impossible* to comprehend. And the inner workings of such psychotic behavior can't be trivial.

Writer Leonardo Sciascia* offers a disconcerting explanation. He compares Ettore to Stendhal, who dreaded his precocity so strongly that he squandered as much time and talent as possible in a vain attempt to "delay" his genius. "Like Stendhal, Majorana tries to avoid accomplishing what he has to accomplish, what he can't help accomplishing," Sciascia writes in his book on Ettore. The Boys were keen in their pursuits, had unbeatable enthusiasm and stamina, but they were "seeking"; Ettore, in contrast, just "found." But such miracles came with a price. Ettore must have sensed death and self-destruction in each of his revelations. They were so extreme that they became sinister, malignant. In this mental climate, throwing away Nobel Prize–worthy work might actually have been an act of self-preservation, Sciascia suggests. So "when Heisenberg's theory is eventually acknowledged and publicized, not only does Majorana refrain from lamenting . . .

---

* Renowned for his literary work on the Mafia and the polemics it engendered. I strongly recommend *To Each His Own* and *The Day of the Owl*.

but conceives for the German physicist a feeling of sincere admiration (based on self-esteem) and gratitude (based on dread). Heisenberg represents for him an unknown friend—someone who without knowing him, has saved him from disaster, enabled him to avoid a sacrifice."

I like Sciascia's affinity for poetic license. How else could one ever explain utter madness?

Back at Via Etnea in Catania, I'd been told I should meet Ettore Majorana, because he now kept the private documents found in Dorina's bedside safe. And so I go to Rome, where Ettore's scientific dramas originally unfolded. The building at Via Panisperna is still there, sadly no longer housing the Physics Department. I have to go instead to the new University City, not far off. That's where Ettore has his office.

SEVEN

Meet Ettore Majorana

✿ ✿ ✿ ✿ ✿ ✿ ✿ ✿

I can't help feeling an odd sensation about sending an e-mail to Ettore Majorana. Such an anachronism. . . . Yet I do e-mail him, and we make an appointment to meet at his office in Rome.

On that morning, the traffic in Rome is the habitual nightmare. Having failed to foresee this, I arrive late. Not *too* late, but I'm striding when I reach the university complex just past the Scuola di Guerra Aerea (the "aerial war" school), a reminder of Italy's bellicose recent past. All around me there's Mussolinic grandeur, imposing marbles, some fake, others not, emulating the style of imperial Rome in a Disneyland of Neros, Caligulas, and bad taste. It feels like I've dropped into a wormhole and been shoveled into the past. Sending an e-mail to Ettore, then, does make some sense.

I rush past the Dipartamento di Igiene (Department of Hygiene) before finally finding the physics building, named after Guglielmo Marconi. Two doors permit entrance into the world of Roman physics: one with Experimental Physics engraved above it, the other with Superior Physics. I smile, considering just how to break the news to my experimentalist friends that they do "inferior" physics. It was here that the Via Panisperna Institute moved in 1937.

A sharp left at the entrance takes me to the Aula ("classroom") Ettore Majorana, where a bunch of young things are making a racket while they wait for a lecture. A door to its side opens into a long corridor. At the end of the corridor, a couple of labyrinthine turns finally lead to his office. Ettore Majorana's office. It says so on the door.

I knock and a middle-aged man appears. He has sharp eyes, dark olive skin, and gray hair. He's polite and very shy. We chat for a while about the state of the art. The experimental detection of gravitational waves is his current interest. As a theorist, I'm always amazed by the ingenuity of experimentalists, as they push technology to the point where it still doesn't properly exist, trying to measure what can't even be detected yet. This is no inferior physics.

We talk at length about the vicissitudes of gravitational-wave detection, but I must eventually fail to disguise that I'm not there to discuss physics. I want to talk human tragedy, the revulsions of the soul, the whys and hows of Ettore's fugue. He lowers his voice, having already shut the door to ensure that no one overhears us.

"So, you want to talk about Ettore Majo . . ."

He breaks the phrase midflow.

" . . . you want to talk about *the other* Ettore Majorana," a gesture emphasizes this detail, "the other, the *important* one?"

He's smiling: It's a joke, and he's enjoying it. There's no bitterness in his voice. So we talk about his eponymous uncle. In English, fortunately! For once the Italian language is spared my usual slaughter.

It must have been tough for Ettore Majorana Jr. to become a physicist. Einstein had a son—Hans—who was a scientist. Imagine what it must have been like for him to go to conferences and write papers. Hans Einstein was a professor of hydraulics at Berkeley University. He was brilliant. But he was no Einstein.

Ettore Majorana Jr. is no Ettore Majorana Sr., but he's an excellent experimentalist. How annoying it must be to have your colleagues expect you to be an outstanding neutrino theorist when you're not even trying to be one. But I feel Ettore Jr. has overcome all that and that he's happy to scorn superior physics to instead play a major role in what could be the next big discovery.

Ettore's English is excellent, but it's obvious that he's mainly used it in a scientific setting. Now—talking about the inner troubles of his uncle—he transplants his technical vocabulary into the alien context of feelings and emotions. The result is exquisite when it's not outright poetic: We do an "inverse reconstruction" of Ettore's soul, discuss a "computer program called suicide," and other such extravagances. Occasionally—when referring to people out of his favor—he can't help slipping into Italian: *"Va fan culo!"* I'd have labeled him sad, from his manner, were it not for these outbursts, which are as hilarious as they are unexpected. I wonder if his uncle's reputation for sadness resulted from a similar misunderstanding.

Like Fabio, Ettore Jr. is the son of Ettore's brother Luciano and Signora Nunni Cirino. In fact, he's the second Ettore the signora gave birth to: The first was born on March 30, 1960 (for those who like coincidences), but unfortunately died the next day. Unlike Fabio, Ettore Jr. lived in Rome for most of his life, lodging with Ettore's younger sister, Maria. His emotional attachment to Maria is obvious: "She was a mother to me. A second mother, since she could never be the first one. But she was a mother for my feelings." He provides unique insights on Maria: "She was very sensitive and affectionate. She translated French poetry in her sleep." And he paints a horrifying picture of the plight of an intelligent woman in the Italy of her time: "When Maria was eighteen, she told the family that she wanted to go to university and do a degree. At once they all started laughing, as if it was a good joke." Was it a nasty put down? "No, no, they laughed *with love*. But still they laughed. And so she never raised the subject again."*

I've been asked if I would have liked Ettore Majorana had I met him personally. All I can say is that I certainly liked meeting Ettore Majorana Jr; and in time, we became friends. I also liked his opinions:

"We live in a world that needs myths onto which we project our hopes, sometimes simply our self-importance. Majorana is a perfect target for mythmongers because we don't know what happened to him. More than not knowing, we can*not* know what happened to him. Why don't we just accept indeterminacy when we find it? We do it in mathematics, why don't we do it with people? An inverse reconstruction is impossible, and the answer to the puzzle is not a solution, but the space of all possible solutions. Just like in mathematics."

Despite the technical language, he couldn't have put it better. His views on Ettore are "nihilistic." It doesn't mean that he doesn't care—quite the contrary. But he's prepared to take an "indeterminate problem" for what it is, and let Ettore's mystery be.

---

* To put things in perspective, Senator Corbino's mother was illiterate. Being a very intelligent woman, she taught herself to read and write at the age of fifty, so that she could find out more about the terrible earthquake in Messina, where her son nearly died.

Although Ettore Jr. never voices a theory, he performs a strange translation of something he'd heard Maria say many years ago (she died in 1997):

"Consider a computer program called Suicide. You press return, and it runs. But something happens during the flow of commands and the program finds an 'interrupt.' Something we don't know caused it, maybe it was raining on that night. The program goes to another line of commands, and it never executes its original purpose, other routines are executed instead; suicide as initially planned is forgone. But then these routines find themselves, quite by accident, at the same point in phase-space that triggered the suicide program to start running. He was on a boat again the next night and perhaps it wasn't raining. So the suicide program is rerun. And this time it does it."

Odd metaphor. But I've often wondered whether it was circumstantial that Ettore didn't jump ship from Naples to Palermo—and then possibly did on the way back. But we simply don't know, as Ettore Jr. emphasizes. And this isn't his theory: He refuses to have one. It's a rendition, in his scientific English, of Maria's theory which I had already heard voiced in a TV documentary, in which she talked about "the power of the *alba* [dawn], that terrible time of day for suicidal people."

"One thing is for sure," Ettore Jr. says, "Suicide or not, he wanted to disappear. To break away from us, the normal ones. And we have to respect that, whatever he did. But now everyone wants to have his own Ettore Majorana. Everyone wants to use him for his own purposes, to further his cause. Why don't we just leave him alone? We should not only respect that he broke away from us, but that he didn't want us to use him."

Our chat gradually moves to Ettore's relations with his family and colleagues. What everyone tries to hide is that he couldn't stand the Via Panisperna Boys, who in turn feared and hated him. Why was he so convoluted? Ettore Jr. suggests a possibility:

"Life on a practical level is a compromise between complementary sets of parameters. Colleagues, family, the self. They all pull in different directions, and you have to maximize some optimization function combining them all. And he was very bad at that: at the business of life.

"His mind was very strong, full of tools. But only tools that allowed him to solve mathematical problems, not the problems of life. In that respect, he was incapable of handling the world. To find a family, to deal with colleagues, to have a career:

in all that, Majorana was handicapped. And eventually life on a practical level caught up with him."

But surely the family must have realized something was badly wrong when Ettore locked himself in his room in 1933 and stayed there for four years. Again, Ettore Jr. voices Maria's views: Of course they knew something was wrong, but no one mentioned it. They just let him be. In the Majorana family (then, as now, he adds sotto voce) no one asks deep questions.

At one stage in the conversation, he stops abruptly, dismisses with a gesture everything we've been considering, and very slowly says the following:

"There is only one thing that I'm sure would have made all the difference in the Majorana affair."

He pauses for dramatic effect.

"Love! Love would have changed everything. The extremes in his personality would have been capped. They'd be low-pass filtered, wouldn't ever manifest themselves. But he never found love. He never found a woman.

"As sure as I'm Ettore Majorana," he smirks at his little pun, "I'm certain that was his real misfortune."

EIGHT

Boys Will Be Boys

✿   ✿   ✿   ✿   ✿   ✿   ✿   ✿

In a beautiful part of town, not far from where I met Ettore, lies Via Panisperna. The building that housed the Via Panisperna Institute has itself become fabled. It was once on the grounds of two convents belonging to the papal state, but when the state was absorbed by the Kingdom of Italy in 1870, the property reverted to the king (as did everything else but the Vatican). After the University of Rome was created, the two convents were donated to become the departments of physics and chemistry.

The chemists performed minimal conversions, turning cloisters into labs and cells into classrooms. The physicists chose to demolish and start anew. Unfortunately, the nuns who lived in the convent took strong exception and refused to leave. Offers of money and other terrestrial enticements failed to produce the desired effect and soon gave way to blackmail and open threats. Still, the religious didn't budge from what had been their seat for centuries.

As a testament to the conflict between science and religion, an infantry detachment was enlisted to displace the nuns and pave the way for progress. Some claim that the building has been cursed ever since—an interesting thought, considering that the fission of uranium was first achieved there.

Nowadays, a persistent rumor maintains that the basement of the distinguished edifice houses a seedy nightclub, but this misconception results from a modern renumbering of the street. In fact, the building is part of a complex belonging to

A view of the modern Via Panisperna.

The building that
housed the Via
Panisperna Institute
(now part of the
Ministry of the
Interior).

the Ministero dell' Interno (Ministry of the Interior), and apparently houses a
high-security jail. This didn't stop me from sneaking in one Saturday afternoon,
while the guards were dutifully watching football on their TV monitors.* As I wandered around unchallenged, I caught glimpses of the idyllic environment gifted
to the Boys.

---

* Lazio versus Roma, for those who know what I mean.

The Boys loved their workplace. According to Segrè, "the location of the building in a small park on a hill near central Rome was convenient and beautiful at the same time. The garden, landscaped with palm trees and bamboo thickets, with its prevailing silence (except at dusk, when gatherings of sparrows populated the greenery) made the institute a most peaceful and attractive place of study. I believe that everybody who ever worked there kept an affectionate regard for the old place."*

There were three floors and a basement. The first floor was used for teaching, and the second for research. The third floor was the official residence of Corbino, who lived there with his family according to custom, amid tiled domed ceilings and ample rooms. The back garden, with its pond full of goldfish, was meant for Corbino's private use, but the Boys overran it: Rasetti used it to rear salamanders; races with candle-powered toy boats gave rise to more bets among the Boys; outrageous nuclear-physics experiments were performed there. The basement may have been the most important of all—it harbored a secret, a treasure more valuable than gold.

The research floor was sharply divided between north and south wings, the former being occupied by Corbino's archenemy Professor Antonino Lo Surdo. It was in the south wing that the Boys practiced their trade. It was a bit cramped, as they had to share it further with Professor Giulio Cesare Trabacchi from the physics arm of the Department of Public Health. But Professor Trabacchi was a generous soul who lent the Boys equipment, chemical products, and even staff— like his nuclear chemist Oscar D'Agostino, without whom the Boys' main contributions to physics wouldn't have been possible. With eternal gratitude, the Boys nicknamed the professor the Divine Providence. Professor Trabacchi, incidentally, owned the key to the safe in the basement, where the dark treasure of Via Panisperna was kept.

But if the Via Panisperna Institute benefitted from a serene location, it was also the scene of much strife. Even the relations between the Boys, and between Fermi and

---

* All material and quotes in this chapter come from Laura Fermi and Emilio Segrè books (see References at the end).

Corbino, could be rather strained. Tempers soared, quarrels developed. The hole that Segrè—the Basilisk—made in their meeting table is witness to how edgy and personal science became there. More generally, as in most Italian academic institutions, there was a state of civil war between various factions within the institute.

For example, a conflict raged between Senator Corbino and the other senior physicist at Via Panisperna, Antonino Lo Surdo. Like Corbino, Lo Surdo was Sicilian: They'd both taught at the University of Messina before transferring to Rome. A feud had started, with Corbino skillfully winning every battle. When Fermi was appointed to Rome in 1927, as Corbino's first salvo to establish the Boys, Lo Surdo had virulently opposed it. When he lost the fight, he felt personally affronted. Lo Surdo thus turned his venom against the Boys, who predictably retaliated with the cruel and merciless humor of youth.

They spread rumors that Professor Lo Surdo had a case of "evil eye" so lethal that "even mentioning his [real] name brought bad luck." It seems that Lo Surdo had witnessed a terrible Italian naval accident in which three hundred people drowned, a man had dropped dead inexplicably on a streetcar platform the moment Lo Surdo stepped onto it, plus a number of similar incidents. Thus, to prevent disaster, the Boys posited that he should be referred to exclusively by a nickname. Due to the location of his office in the north wing, "Mr. North" was chosen. The poor man then became the butt of an "evil eye" running joke; e.g., "the best hydrogen tubes exploded for no reason other than that Mr. North minutes before had told the student working with them: 'Be careful, it might explode.'" Now consider that Sicilians are invariably superstitious. The Boys must have driven the professor mad.

When Mussolini created the Royal Academy of Italy in 1929, Lo Surdo—a rabid Fascist—nourished high hopes of being included in the first batch of appointments. His expectations, however, were brutally cut short by Corbino's pervasive political influence. According to Laura Fermi, when the announcement was made, "Amaldi ran to the North side of the building shouting: 'The first academicians have been named! There is also one physicist . . . ' Mr. North's bulging eyes popped out further behind his thick lenses. His cheeks were aflame. 'Fermi!' Amaldi exclaimed with candid joy, then turned away as fast as he had come, but not before he saw Mr. North's face turning purple."

Like Corbino, Lo Surdo had been in Messina on December 29, 1909, the day a massive earthquake and tidal wave killed a third of the population. Mr. North had lost all his family, as well as the girl he hoped to marry; whereas Senator Corbino had miraculously escaped unharmed with all his relatives. Afterward, Mr. North led a lonely life, never marrying or making any friends. It's telling that his research shifted from atomic physics to seismology. It seems that his bitterness towards the senator derived from this tragedy, which he perceived as an unfairness of fate. Is his attitude more palatable in this light?

Not to the Boys, that much is for sure. Their "evil eye" stories, applied to someone so unlucky, were in very poor taste, to say the least.

By around 1930, the Via Panisperna Institute had become well-established in Italy and abroad. Its staff grew as the institute attracted a second generation of Boys from Pisa, Florence, and Turin, all in search of a slice of glory. Giancarlo Wick and Bruno Pontecorvo would be the most distinguished of these latter-day Boys. There was also a profusion of largely forgotten minor Boys, and even—God spare us!— failures. One of the earliest Mark II Boys who did stand out was Giovanni Gentile Jr., son of the eponymous patriarch of Fascist culture. Gentile became Ettore's best friend and ally at the institute.

As Via Panisperna's reputation spread abroad, a stream of illustrious foreign visitors also began to pass through. Some stayed for long spells, writing landmark papers and performing historical experiments. Names like Bethe, Block, Peierls, Placzek, and Teller lent the institute a cosmopolitan tinge; foreign languages now echoed in its corridors. These famous visitors were attracted by the institute's rising fame, but some were also escaping Germany's developing political situation. Many were German Jews en route to permanent resettlement in the United States. There were also several American visitors, of whom Eugene Feenberg was Ettore's favorite: "Their mutual attraction manifest[ed] itself in their sitting in the library facing each other in silence because they knew no common language."

Being on the world map must have been of little import to Ettore, who remained a lone wolf, as before. It must have been inspiring for Fermi, though, who had previously felt "somewhat isolated because only Majorana (who was rather inaccessible anyway) could speak with him about theory on an equal footing," as Segrè commented. During his time at Via Panisperna, Ettore wrote only one collaborative paper, with his friend Giovanni Gentile. To the other Boys, engaged in permanent collaborative work, he must have seemed like a black hole—a human neutrino. Typical testimony: "I avoided talking physics with Ettore because anything I could have told him would have been insignificant for him. As it happened to me with Pauli later, Ettore must have thought that it was more accessible for me and less banal for him, to communicate, for example, how good it was that he'd been born after Michelangelo and Beethoven."*

But with Giovanni Gentile things were different—another example of Ettore's duplicity between extreme warmth to his friends and zero-Kelvin frost to everyone else. Ettore and Gentile were the same age almost to the day and had met via the Casina delle Rose group. Gentile had first joined the Boys as an unpaid assistant to Corbino in 1928, but had interrupted his studies to do his military service like a good Fascist. He then rejoined for good until the strange events of 1937. Like Ettore, Gentile was Sicilian. They became very close.

And just as with Gastone Piqué before, Ettore's letters to Gentile provide our best insights into his attitudes on life, science, and his own work. Self-deprecation, lack of seriousness and sarcasm are the main flavors. At the age of twenty-three, after dryly describing "the Pope's" recent work in his letter, Ettore adds:

> As for myself I do nothing sensible. That is, I study group theory with the firm intention of learning it, similar in this to that Dostoyevsky character who started one day to set aside his small change fully persuaded that soon he'd be rich like Rothschild.

Or a year later, while Gentile was visiting Heisenberg in Germany:

> Rome doesn't present to me, like Leipzig to you, the attraction of novelty, neither does it feed under cloudy skies vast masses of grave thoughts. Rare are here the causes to meditate; and even rarer those who engage in it. And still they say that this is the blessed land of God! The thousand German luminaries, shining like beacons in the snow, compete without success with our burning Sun as it graces the poor mortals with its light and warmth. . . . I say this not to diminish our respect for such a tenacious race, but so that

---

* This testimony comes from a 1984 letter from Gilberto Bernardini to Erasmo Recami. Bernardini was part of the Florence group, set up by Fermi's friend Enrico Persico. The two groups kept a close collaboration.

it won't be uncritically that we voice our interest in their works and also in their sterile attempts; because that which fails elsewhere is destined to triumph under our more friendly sky. Moving on to what might be closer to our interests, I can report to you that the rhythm of scientific activity at the Institute is in decline due to the habitual effect of the summer heat as well as the upcoming Pontifical departure, which will take place as you know on June 7. One of these days I shall see Pirandello's "As you desire me."

Fermi made no effort to polish his articles and, according to Segrè, wrote in a "flat careless language." Ettore's letters, in their eloquent style if nothing else, emphasize the abyss that separated him from the Boys.

The period between 1930 and 1933 was Ettore's most prolific: The small change he so disparagingly set aside led him to a colossal masterpiece, which I like to call his unfinished symphony. His work to a large degree prefigures modern attempts at grand unification, as I shall explain. But he was far from happy. The fact is that not all was a bed of roses at Via Panisperna. Indeed the relations between Ettore and the Boys were fraught with the highest-voltage tension.

Most obviously, Ettore didn't like Rasetti and Fermi's brand of humor. Their pranks were deeply puerile (stink bombs, etc.), their games childish to the point of ridiculousness, their humor educational, patronizing, wholly unaware of the human dimension. Reciprocally, one wonders how Ettore's sophisticated "British" sense of humor was received. If it registered at all, I'm sure it caused more hurt than mirth.

Rasetti and Fermi also had an unbearable way with women, always lecturing and then quizzing them on butterfly species, the exchange rate of Brazilian reals, the death dates of all the kings of England, and other inane trivia.* "Those two

---

* Sample: The boiling point of oil (figure supplied) is higher than the melting point of the metal used to manufacture a normal pan (figure supplied, too). How do you explain that deep frying doesn't melt the pan? Answer: When you deep fry, the oil never boils. It's the water contained inside the food you're cooking that does. Fascinating.

or alive, in the cat's case) if an observer entered the picture. These views are solipsistic. I doubt Fermi even knew what the word *solipsistic* meant. He was a great problem solver, but his strengths finished there. Unsurprisingly, in a generation of philosopher physicists, Pauli called Fermi a "quantum engineer."

In her book on Fermi, Laura reports that her husband once set out to prove to her that light was an electromagnetic wave. Laura complained, "No! You proved that you can obtain two equal numbers. But now you talk about the equality of two things: You can't do that! Besides two equal things need not be the same thing." Which I think is quite a good point. But Enrico was furious.

In 1942, Fermi didn't think twice about testing the first nuclear reactor in the middle of Chicago. It's nerve-racking to read reports of the first nuclear pile going critical, kept under control by Fermi doing calculations with a slide rule and shouting instructions across the room to the guy controlling the moderator. This first "controlled" nuclear reactor could have ended up like Chernobyl in the middle of a large town! But like ontology, ethics was a branch of philosophy that didn't interest Fermi.

A schism was therefore developing at Via Panisperna as 1933 approached, a rupture mimicking the rift between Ettore and Fermi. You can feel it in the testimonies provided by Ettore's surviving mates in latter-day TV interviews. If you were to hear me reading this page instead of reading it yourself, you'd most likely gain a better sense of who I am. The voice is the music of the soul. Ettore's friends, Gastone Piqué, Giancarlo Wick, and Giuseppe Occhialini, sound affectionate, full of human confusion, endearing in their obvious love for a fellow soul they tried, but failed, to understand. Fermi's side, Amaldi and Segrè, come through terribly— their voices fast paced and toneless, stiff with arrogance and authority, devoid of doubt. Their voices sound like robots programmed for omniscience. There are no uncertainties in their statements. Not even quantum mechanics is like that.

A mere personality clash . . . had history not intervened.

One should be aware that while all these events were taking place a powerful political stew had just begun to simmer. Since 1922, Italy had become a Fascist country, indeed the word was coined there (after *fascio*, Italian for "league" or "unity"). The founder was Benito Mussolini, and he carried all the marks of what was to come: authoritarian, vain, ruthless, populist, and not very intelligent. Mussolini's

IL GOVERNO FAſCISTA
MI HA RIDATO LA MIA DIGNITA'
DI LAVORATORE E DI ITALIANO

Typical Fascist poster: "The Fascist government has given me back my dignity as a worker and as an Italian."

Italy became an early dictatorship (together with the Salazar regime in Portugal), providing the loose mold for Hitler and Franco years later.

But the Fascist movement initially seemed more innocuous than later developments would prove. In 1922 the Fascists marched on Rome, carrying nothing but their enthusiasm for the glorious Italy of a remote past. As Senator Corbino put it to Fermi on the day the Blackshirts arrived in Rome, "They have no arms; there will be a massacre. What a pity! So many young men will die who were only searching for an ideal to worship and found none better.* "But there was no massacre, the king played along, and with his trademark astuteness Senator Corbino predicted that Italy would become a dictatorship under Mussolini.

Yet fascism didn't degenerate into dictatorship at once, and even after the withdrawal of civil liberties (for example, multiparty elections and freedom of the press), many believed that these were temporary measures at a time of strife. In addition, fascism brought economic relief to many poor Italians. In a deeply Catholic country, where the church and state had been in conflict since the nineteenth-century reunification, fascism achieved reconciliation. It stirred national pride in its obsession with Italy's past glories. When Umberto Nobile succeeded in overflying the North Pole aboard a dirigible in 1926, it meant a large dose of much needed self-respect for Italy.

So much so that two years later, when Nobile crashed while attempting a landing on the North Pole (leaving many dead, including pioneer Roald Amundsen, who commanded the rescue party), a nasty surprise awaited him back home. For all his efforts and risks Nobile was publicly humiliated, discharged from the army, and stripped of all his medals, narrowly escaping prison. Fascism didn't tolerate failure.

Fascism was idealistic, in its strictly philosophical sense. Giovanni Gentile Sr., the father of Ettore's best friend at Via Panisperna, was responsible for wedding Fascist doctrine to idealistic philosophy. The resulting synthesis is "condensed" in the Italian encyclopedia, a ridiculously expensive mastodon, so large it was never

---

* The quote comes from Laura Fermi's book, *Atoms in the Family*.

reedited. On one point you cannot fault idealism: It is logically inconsistent with racism. If the body and the soul are detached, and if the soul is the essence of the human being, then a genetic label is meaningless, and so discriminating against blacks or Jews is impossible.

Umberto Nobile's blimp, *Norge.*

But of course no one thought about the dangers of racism in the early days of fascism, or indeed well into the mid-1930s. Racism and anti-Semitism were not an issue in Italy, and in fact Germany and Italy were traditional enemies. In 1934, Mussolini moved troops to defend the Austrian border against an early German threat, after a Nazi putsch. The Germans were duly intimidated by the Italian army, at that stage more powerful than theirs. Hitler was widely seen as a joke in Italy, with nothing in common with Mussolini. There seemed to be no chance of any alliance between the two. Why, then, worry about German-bred racism and anti-Semitism in Fascist Italy? In 1932, Mussolini stated in a speech that "from happy intermixing of peoples results the strength and the beauty of a nation."

The political turmoil had barely begun.

Yet by 1932, Ettore's downward spiral was already well in motion. It would lock him in a black hole of despondency and dread from which he'd be unable to emerge for years. There were so many forces at play that it would be foolish to place undue emphasis on any one of them. They all dragged him down into the abyss in a cacophony of gargoyles: Fascist and Nazi politics, the burnt-baby scandal, poor health, scientific troubles. . . . In no particular order, let's begin with a particularly heavy stone that looped around Ettore's neck as he swam happily in the sea of science, finishing off what he regarded as his masterpiece. Meet Professor Paul Adrian Maurice Dirac, the autistic physicist. And unwittingly a major force in Ettore's downfall.

ELEVEN

Creation and Annihilation

❆ ❆ ❆ ❆ ❆ ❆ ❆ ❆

Sometimes I think that despite their scientific differences, Ettore and Dirac would have been the best of friends. They found themselves at opposite ends of a scientific dispute, yet their mathematical minds worked in similar ways, and their personalities had much in common. But to the best of my knowledge they never met. What a pity: It would have made for some first-class anecdotes.

Dirac was Ettore without his gregarious side, to the point of borderline if not outright autism. He seldom spoke, but when he did the outcome was minimalistic, logically crafted insanity. He was incapable of *not* thinking literally. At a conference someone got up and said, "Professor Dirac, I don't understand how you derived the formula on the top right-hand corner of the blackboard." To this he responded with silence. When the chair inquired why he didn't answer the question he replied, "That was not a question, it was a statement."

Perhaps Dirac's genius derived precisely from the savant streak present in some autistics. When a Russian colleague gifted him Dostoyevsky's *Crime and Punishment* his opinion was "It's nice, but in one of the Chapters the author made a mistake. He describes the Sun rising twice on the same day." Was this uncanny ability to discern irrelevant detail an autistic-savant trait? It certainly led him to new mathematics where others had failed. But on the very few occasions when he commented on his own work, he claimed to be guided by "mathematical beauty," a concept he considered superior even to simplicity. "It is more important to have beauty in one's equations than to have them fit experiment"—the ultimate victory of Platonism.

Dirac was even more taciturn than Ettore, and his inability to make small talk was legendary (that his wife, Manci, never shut up was apparently unrelated). A visitor to Cambridge narrated, "The weather outside was very bad, and since in England it's always respectable to start a conversation with the weather I said to Dirac, 'It's very windy, Professor.' He said nothing at all, and a few seconds later got up and left. I was mortified, as I thought that I had somehow offended him. He went to the door, opened it, looked out, came back, sat down and said, 'Yes.'"*

Unfortunately, I have the dreadful feeling that none of this outlandish behavior is actually funny; that it was involuntary, something he couldn't help, and caused him much suffering. Dirac was severely abused as a child. If Ettore was traumatized by his mother, Dirac endured an authoritarian and sadistic father. "I never knew love or affection when I was a child," he'd say cryptically in his old age. His brother committed suicide while still very young, and Dirac blamed his father. When Dirac won the Nobel Prize in 1933, he invited only his mother to his lecture. The extent of the abuse to which he was subjected has never been fully investigated. For someone so quiet and private, it's harrowing to hear that Dirac once publicly said that his father was the only person he ever loathed.

Paul Dirac was born in Bristol, England, on August 8, 1902, and, like Ettore, first trained as an engineer. His aversion to experimental work, however, soon led him to switch to mathematics. While Ettore was attracted to Via Panisperna, Dirac fled to Cambridge, that ivory tower of lunacy, where his behavior passed largely unnoticed. In the early 1990s, when I was a fellow at what had been Dirac's college, he was still fondly remembered (unlike Manci, who terrified generations of misogynist

---

* John Moffat's book of memoirs *In the Company of Giants* contains a number of such delicious stories. I can also heartily recommend the recently published Dirac biography *The Strangest Man*, by Graham Farmelo. Having spent twenty years berating my British colleagues for the absence of a biography for their twentieth-century scientific exponent, I am happy to say that the wait was worth it. The material in this chapter comes from this source.

academics). At the high table, "the Dirac" was still in use as a unit of "verbal output," similar in nature to the carat or the microgram for weight.

In 1926 Dirac finished his Cambridge University PhD degree, with a thesis titled "Quantum Mechanics," the equivalent of titling a book *Poetry* or *Science Fiction*. But it's a befitting title: It's where the foundations were laid down for what had been a fragmentary hotchpotch. Overnight, the twenty-five-year-old Dirac became the leading authority in quantum theory. Dirac's formulation of quantum mechanics led to all the modern offshoots, including attempts to quantize gravity.

After completing his PhD, Dirac embarked on an even more ambitious project: "relativistic quantum mechanics." In its initial formulation, quantum mechanics was only applicable to particles moving much slower than the speed of light. Einstein's theory of relativity gives a radically different picture of the universe when things move fast. Clocks tick more slowly, distances shrink, objects become "heavier" and more difficult to accelerate. A relativistic quantum theory that unified quantum mechanics and special relativity was lacking.

No one knew how to describe quantum phenomena for particles moving close to or at the speed of light. Dirac set about filling the gap. The result was Dirac's theory of the electron. And most infamously, the antiworld.

Dirac's discovery was made purely via mathematics, and at first it didn't conform to experiment *at all*. Had it not been for his obsession with "mathematical beauty" (to the detriment of experiments if need be), I suspect that Dirac would have given up and the antiworld been found only much later. He worked with formulae and abstract concepts, conciliating Einstein's theory of special relativity and quantum mechanics in his mind, setting up a mathematics embodying both. It took him several months of trial and error in 1927 to hatch his theory. But when he finally devised an equation describing a quantum-mechanical particle (with spin and all the quantum trimmings) moving close to the speed of light, he got a surprise. His equation—nowadays called Dirac's equation—described not one particle, but two.

Somehow the mathematics doubled the number of particles in the theory once quantum mechanics and special relativity were married. Taking the electron as an example, Dirac's equation contained solutions with the correct mass, spin, and

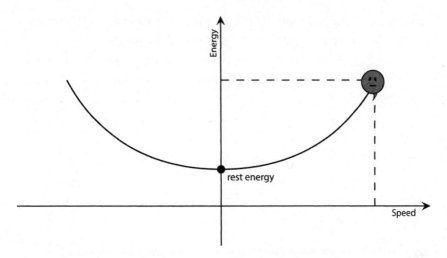

Figure 11.1: The positive energy "bowl." At zero speed, the electron has a rest energy E = mc², associated with its mass at rest, m. As the speed increases (for both positive and negative speeds), its energy increases. (The bowl extends upward forever and is actually infinite.)

charge and, crucially, with positive energies (see Figure 11.1). But there was also a whole world of identical electrons, with the same mass and spin but *negative* energies (see Figure 11.2).

Physicists don't like negative energies and for good reason. If you accelerate a negative-energy particle you gain rather than spend energy. Sounds like a cool solution to the world's energy crisis, but things just don't work that way. Furthermore, I'm not sure we'd like the free ride: We wouldn't be able to stop the process, and it would end in apocalypse. In the same way that all matter rolls to its minimal-energy state, electrons with negative energies would indulge in a runaway behavior, where they would accelerate more and more (causing their energy to become more and more negative), meanwhile flooding their environment with positive energy, in a spiraling, uncontrollable mess.

The solution to the problem found by Dirac is ingenious. Nowadays it's considered heuristic, and there are more mathematically rigorous alternatives, but it's still an intuitive depiction.* Dirac cured the instabilities associated with these

---

* Dirac's heuristics are also a technically correct account of some of the physics that rules, for example, the electronics in your mobile phone.

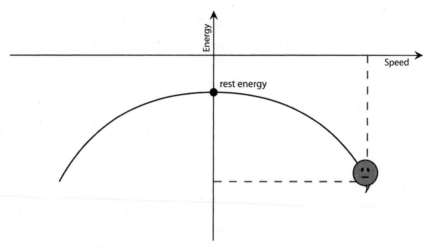

Figure 11.2: Dirac's inverted energy bowl, servicing negative-energy electrons. The rest energy of such electrons is negative, and their energy decreases (it becomes more negative) as their speed increases. It therefore costs nothing (indeed energy is gained) to accelerate a negative-energy electron. A runaway instability follows.

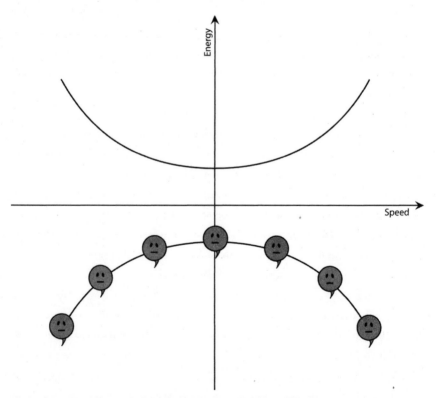

Figure 11.3: Dirac's cure to the negative-energy instability: Fill all the negative energy states with particles so that no state is left free and define such a "sea of electrons" as "nothing." Negative-energy electrons now cannot run away down the bowl, because the bowl is entirely full.

negative energies with the following recipe: Fill all the negative-energy states with particles, so that no state is left free. And call *that* "nothing," or the vacuum (see Figure 11.3).

If you accept this "redefinition" of the vacuum, stability is accomplished. Negative energy particles can't run away into even more negative states (shedding an overdose of positive energy into the rest of the world), because all such states are already filled. Neither can positive-energy particles jump into negative-energy states. The universe becomes stable because the negative-energy states are all filled by default. Crucial to Dirac's argument is the realization that you can't put two particles into the same state (at least not for fermions, precisely according to the principle Fermi failed to grasp ahead of Pauli).

But if this "Dirac sea" of negative-energy particles is declared to be the vacuum, then what is "something" that exists, "something" that isn't part of "nothingness"? One only measures differences between things, so reality—existence—should be what's new *with respect to* Dirac's sea. Such as an electron with positive energy— a regular electron, negatively charged as usual (see Figure 11.4). *Or* a hole in the Dirac sea: the failure of one of the negative-energy states to be filled by an electron. Such a hole would be perceived as a deficiency in negative electric charge and in negative energy. It would therefore be seen as a particle with *positive* energy and *positive* charge (see Figure 11.5 and Figure 11.6). Less debt is equivalent to more money: Likewise, less negative energy and charge is equivalent to more positive energy and charge. Thus, the prediction of the antielectron, or positron: a particle with the same mass and spin as the electron, but opposite electrical charge—and with positive energy, of course. The antiworld was unveiled.

I'm told that on cruises, if the ship's engines are turned off in the middle of the night, everyone wakes up startled as if by the loudest of dins. The engine's hum has become the vacuum. Its absence is now perceived as a loud, sleep interrupting, antinoise. A similar argument willed the positron into existence.

In fact, the electron is not alone and any charged particle, by the same token, must have a corresponding antiparticle with the same mass and spin but opposite charge. There is an antiproton, whose experimental discovery earned Segrè the

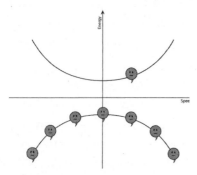 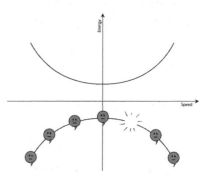

| Figure 11.4: A positive-energy electron hovering above Dirac's negative-energy sea. The sea passes unnoticed (because it's "nothing") and only the positive-energy electron is noted. | Figure 11.5: A hole in Dirac's sea would necessarily be seen as "something," since it constitutes a blemish in the vacuum, or the nothingness. |

Nobel Prize in 1959. The antiproton has a negative charge like an electron but the same mass as a proton. Likewise for all charged particles.*

Our world is primarily made of matter, and antimatter is rare, produced only in physicists' accelerators or by cosmic rays. Just as well. Because something dramatic happens when matter meets antimatter: They annihilate.

Consider Dirac's picture of the positron as a hole in a sea of electrons filling all the negative-energy states. If a regular electron (i.e., an electron with positive energy) meets such a hole it happily jumps into it (things love going downhill). In the process it sheds the large energy difference between the two states in the form of light (typically two photons). At the end we have the hole filled and no positive-energy electron; i.e., we have what we defined as nothing. The electron and positron have *annihilated*, releasing their inner energy in the form of very powerful light (see Figure 11.7). The amount of energy released in these annihilations is enormous. Just plug the electron mass into $E = mc^2$, and you'll find that you wouldn't want to be anywhere nearby. A bomb made of matter and antimatter brought together would be more than a thousand times more powerful than an atomic bomb with the same mass. Except that I'm not sure how such a bomb could be manipulated without detonating immediately.

Conversely, a large amount of energy (say, in the form of two photons) could make an electron in the negative-energy sea jump into a positive-energy state (see Figure 11.8). At the end, with respect to the vacuum, we have a positive-energy

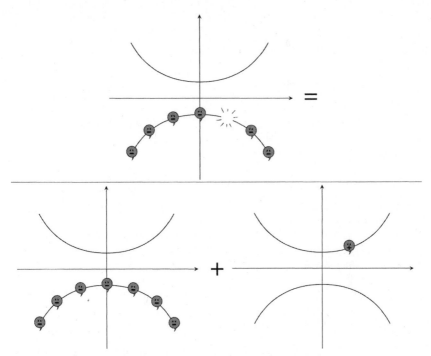

Figure 11.6: A hole in Dirac's sea is equivalent to the sum of a perfectly filled sea and a particle with opposite qualities to that missing in the hole; i.e., a particle with positive energy and charge (notice the "mouth" of the positron). Thus was the positron predicted.

electron plus a negative-energy hole; i.e., an electron plus a positron. Two very energetic particles of light can therefore create a pair of matter and antimatter particles in a process reciprocal to their annihilation. Creation and annihilation are two sides of the same coin—and are the most distinctive hallmark of Dirac's theory.

Nowadays Dirac's theory is lauded as one the greatest achievements of the twentieth century, but no one believed it at first (and Pauli took the piss out of it

---

* For neutral particles the story is more involved. For example, the photon doesn't have an antiphoton, or seen in another way, the photon and the antiphoton are identical. But the neutron does have a very distinct antineutron. As for the neutrino . . .

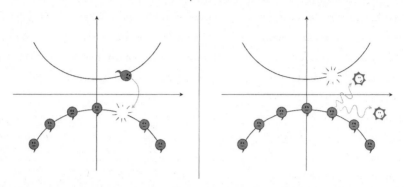

Figure 11.7: An electron-positron annihilation. If a positive-energy electron meets a hole in Dirac's sea (a positron) it may jump into it, releasing the energy difference in the form of two very energetic particles of light. At the end, the hole is filled at the expense of the positive-energy electron. We have, therefore, a perfect Dirac sea, or a vacuum: The two particles have annihilated.

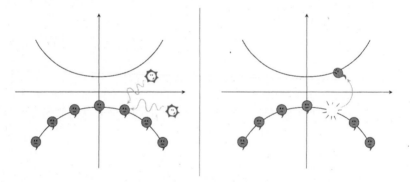

Figure 11.8: The creation of an electron-positron pair. In the converse process to annihilation, two very energetic photons knock a negative-energy electron into the positive band. With respect to the vacuum, we end up with a positive-energy electron plus a negative-energy hole; i.e., an electron plus a positron.

tremendously). In his first paper, Dirac wrongly stated that the antielectron was the proton; Oppenheimer duly pointed out that particles and their corresponding antiparticles must have the same mass. The proton, recall, is 1,836 times "heavier" than the electron.

But Dirac found the mathematics in his theory beautiful, so he stuck to his guns. OK, so his theory didn't describe the electron and the proton: It described the electron and *another* particle with the same mass and spin but opposite charge, which had yet to be seen. Somebody would eventually find the positron, he was sure. Dirac thus started the trend of postulating new, as yet unobserved particles that has continued to this day, when everyone fancies themselves a little Dirac. Thousands of unobserved particles fill the tomes of theoretical physics, waiting to be detected in order to provide the proposer with a suitable Nobel Prize. Too good to be true, their enthusiastic creators seemingly fail to notice.

Of course, the story with Dirac was different, but as Ettore came into the picture there was no evidence for the positron. Ettore wasn't amused. To him it was all utter rubbish: negative energies, particles that didn't exist, and whatnot. His anti-Dirac reaction would lead him to the Majorana neutrino, but Ettore's claim to Nobelization is much more profound: His neutrino is merely the bastard child of much deeper work. What really worried Ettore was what he perceived as the sheer nonsense of Dirac's sea and its negative energies. This opposition to Dirac's theory led him to a masterpiece, "Relativistic Theory of Elementary Particles with Arbitrary Spin" published in 1932: his unfinished symphony.

But then, in the months that followed its publication, a strange set of events was set in motion in Ettore's life.

TWELVE

The Serpent's Egg

By the end of 1932, at the age of twenty-six, Ettore had become a mature scientist in the eyes of the establishment. He'd finished his *libera docenza**; had written groundbreaking papers in atomic, nuclear, and particle physics; and was internationally renowned. A generous grant of 12,000 lire, personally signed by Marconi, was awarded to him (for comparison, a full professor earned about 25,000 lire per year). Ettore was fast becoming famous. He even joined the Fascist party.

For the record, by 1933 everyone at Via Panisperna, with the sole exception of Corbino—who had other strings to his bow—was a party member. It was a mere formality, handy for cutting through red tape and when applying for state jobs or obtaining grants. Fermi joined in 1929; Ettore's Zio Quirino in 1926; Lo Surdo—the only one who was a real Fascist—in a record 1925. Even Giulio Racah from the Florence group (later to emigrate to Israel and found physics there) was a member. Ettore wasn't in the country when the relevant paperwork was signed, suggesting that he sent someone (his brother Salvatore?) to impersonate him. Like

---

* The career of an Italian scientist began with a PhD degree (which Ettore obtained in July 1929, at age twenty-two). The aspiring scientist then became an assistant (*aiuto*) to a professor and eventually, via research and fame, a *libero docente*, or unpaid lecturer. After a few years of unpaid toil, the *libero docente* could finally apply for a tenured chair in a *concorso*, an open competition.

everyone else, he wore the Fascist party pin in his lapel, yet "every first of January he made a bet with a friend that before the end of the year Mussolini would be thrown out. And at the end of the year, at San Silvestre, he'd pay up the lost bet," according to Maria.

Grant in hand, Ettore journeyed to Leipzig in January 1933 to join the Physics Institute. Which meant Heisenberg—his scooper and savior. And then a miracle happened. Ettore's social reticence, his difficulties with dialogue and friendship, completely dissipated. Reading his letters and pooling recollections from others, we find him incongruously upbeat, positive, open to all. "At the Institute I've been received very cordially," Ettore wrote to his family in January 1933. "I had a long conversation with Heisenberg who is a person of extraordinary courtesy and sympathy. I'm in excellent relations with all, particularly with the American Inglis whom I already knew from Rome and who keeps me company and often plays the guide.... The climate is pleasant ... and life isn't expensive. The numerous cafes and nightclub venues are good value, with great music and carnival-style happiness, brimming with people every Saturday evening. . . . The Institute of physics . . . is located at a pleasant location, slightly out of hand, between the cemetery and the madhouse."

He went to the cinema, plunged into the nightlife, found the town comfortable and its people welcoming. Contrary to his habits, he even bothered to publish his work: "I've written an article on nuclear structure that pleased Heisenberg a great deal, even though it corrected his own theory. I'll also publish in German an expanded version of my last article in 'Nuovo Cimento'; in this paper one can find an important mathematical discovery, as it has been confirmed to me...." He's referring to his 1932 masterpiece, notably proud of his achievement, parading it around Leipzig, for once sensitive to praise and recognition.

The miracle was accomplished by none other than Heisenberg, who was then only thirty-one but already a full professor, famous for the discovery of his "uncertainty principle" and the mathematical framework of quantum mechanics. Heisenberg led a hedonistic lifestyle: He drank liberally and was an unabashed playboy; he played the piano exquisitely (if rather extravagantly) and was a top-class dancer; he was a recklessly speedy skier, always going off to the mountains in search of a good time. Heisenberg invariably woke up late and worked only in the afternoon: He claimed he could only do physics if he was enjoying life.

In keeping with his lifestyle and personality, Heisenberg kept things informal and lighthearted at his institute, but without the foolishness of Via Panisperna.

He had a Ping-Pong table installed in the library, and much time was spent playing rather than working. But unlike Fermi, Heisenberg was highly cultured and could also be a deep thinker. His father was a classicist, and Heisenberg had been educated in that tradition, reading Plato and Aristotle's major works in the original Greek. He had strong humanistic leanings and besides cutting-edge physics, philosophical discussions were part of the work schedule at his institute.

Together with Niels Bohr—his mentor and the founding father of quantum mechanics—Heisenberg was concerned with the ontological implications of the new theory. According to the most direct interpretation of quantum theory (to this day controversial), the world can happily live in superpositions of opposites until an observation is made, forcing reality to collapse into a solid certainty. The old philosophical threat of "reality as a projection of an individual mind"—or solipsism—had broken into science. Some of Heisenberg's early works are a mixture of physics and philosophy, addressing the issue of solipsism very closely.

This was precisely the side of quantum mechanics Fermi could never understand, but which appealed so much to Ettore. His discussions with Heisenberg must have covered everything from neutrons to philosophy and literature. He wrote of Heisenberg in his early 1933 letters: "We have became great friends after many a conversation and some chess matches." And "[Heisenberg's] company is irreplaceable for me." So complete was their rapport that they became a threat to everyone else, particularly during seminars, where they ganged up on critics of their strong nuclear-force theory. It seemed that Ettore had found a twin soul. And this probably explains why he became so outgoing.

But the revolution in his life extended beyond his relations with colleagues. He reported home that he was in good health, even though he drank liters of coffee and chain-smoked throughout this period of intellectual ebullience. On his own, without his mother, he dealt well with "life on a practical level," to use Ettore Jr.'s turn of phrase. In his letters to his parents he notes and comments on the financial crisis that followed the Great Depression in America, expertly changing his lire as most appropriate, given the general volatility in the exchange markets. He became a man of the world, and it was through his diligence (by introducing him to an Italian employer) that his sister Rosina's future husband, Werner Schultze, came to Rome. The escalating German political situation, mounting all around him, didn't pass unnoticed to Ettore, either.

The interesting timing of Ettore's sojourn in Germany has probably not escaped you. Barely a week after Ettore's arrival, following a protracted political crisis, Adolf Hitler was nominated chancellor of the Reich. The gears were set in motion. It would have been difficult to ignore what happened next—even at an institute of physics.

On the night of February 27, the Reichstag fire took place, an obvious act of arson that destroyed the German parliament. On the surface, this seemed to be a Communist plot, but it was actually a bit of theater orchestrated by Nazi storm troopers. Like modern terrorism (real or perceived), the fire gave Hitler the perfect excuse to suspend individual freedoms and assert his rule. On March 5, rigged elections gave a majority to the Nazis, who passed the Enabling Act on March 24, transferring full powers to Hitler. All opposition was ruthlessly destroyed in a series of purges that led to the establishment of the first concentration camps. Racism and anti-Semitism became widespread and part of the state ideology. March 31 was declared National Day of Boycott of the Jews, when storm troopers beat up or killed prominent Jews and those who engaged in business with them. On May 10, a large group of Nazi students congregated outside Berlin University to burn 20,000 Jewish books. "These flames not only illuminate the end of an old era, but shed light on the new one," said Goebbels, the

Boycott of the Jews demonstration. In this picture German Jews were made to march carrying signs stating "A good German doesn't buy from the Jews," and the like. The men in uniform are SA.

state ideologue. Jewish composers, and writers like Thomas Mann and Eric Maria Remarque, came under attack. And physics would not avoid the line of fire.

Strong credence was suddenly given to a movement called Deutsche Physik, which for years had attempted to establish the concept of Aryan—as opposed to Jewish—science. The movement was headed by two Nobel laureates, Johannes

Stark and Philipp Lenard, and at best it was a critique of theoretical physics in favor of experimentation, at worst a mere smokescreen of racism and forceful interpretations. Unsurprisingly, their prime target was Albert Einstein: They condemned his theory of relativity as "the great Jewish bluff." Einstein chose at first to ignore rather than respond to their attacks, but by 1933 he was forced to emigrate to the United States.*

Although Heisenberg was a staunch right-wing nationalist with undeniable "Aryan" credentials (he'd been part of the German Youth movement, for example), he wasn't a moron, and all this nonsense against the theory of relativity irritated him. He opposed Deutsche Physik and was soon dangerously high up in their bad books.† A public political argument broke out as to how the scientific interests of the Third Reich could best be served; the Deutsche Physik mob called Heisenberg a "white Jew," providing a sense of the level of the debate. The early squabbles of 1933 would erupt into a full confrontation over the succession of the main professor in Munich (Arnold Sommerfeld), in what would become known in Nazi Germany as the "Heisenberg affair." Heisenberg stood his ground, suffering heavily from his anti–Deutsche Physik stance. Who knows where his mind was while he discussed physics and philosophy with Ettore.

Perhaps so as not to alarm his family, Ettore refrained at first from commenting in his letters on the political incidents. He only mentioned hearing gun reports in Leipzig on the night of the fire, months after the fact. It wasn't until May of 1933 that his letters began to reflect his views on the subject. We will examine these letters in detail, for the role they played in his personal tragedy. Because it was in one of his letters commenting on Nazism that Ettore did something "downright inexcusable" to Emilio Segrè.

* Einstein resigned from the Prussian Academy in 1933; but his resignation was not accepted—just so that the academy could then "expel" him. Nice.

† While all other non-Jewish German scientists were washing their hands of the matter (when not opportunistically using the situation to promote their careers), Heisenberg offered to keep Max Born in office, after the ethnic cleansing of May 6 came into force. Born, a Jew and a close friend of Einstein, contemptuously declined the "favor."

As March 1933 approached and German academia broke up for Easter, Ettore moved on to Copenhagen. He liked the town a lot less than Leipzig but was still in high spirits as he joined the Niels Bohr Institute. He commented, tongue in cheek, in a letter to his father: "[Bohr is] the great inspiration of modern physics, now a bit aged and considerably senile. He still passes for a deep thinker; he speaks a jumble of all languages swallowing his words, giving the impression that if he bothered with his elocution other people might actually understand what he's saying. Everyday he writes one word for his new article, which everyone is certain will be of decisive importance."

Unlike relativity, which came to the world in one piece by the grace of the great Albert, quantum mechanics was a collective and multistage effort involving two generations. The early stages were the work of Max Planck, Albert Einstein (in his spare time), Prince Louis de Broglie, and, above all, Niels Bohr. Bohr also acted as a pivot to the second generation and the final formulation of the theory, as proposed by Heisenberg, Schrödinger, Dirac, and Pauli. So productive were those days that there weren't enough Nobel Prizes to go around. Heisenberg won the 1932 Nobel Prize, but lest it might cause offense to his equally deserving competitors, this was only announced in November 1933, at the same time the 1933 Nobel Prize was awarded to Schrödinger and Dirac.

Niels Bohr was a national hero in Denmark and physically "looked as if he had absconded from the captaincy of a herring trawler," in the words of Graham Farmelo. Bohr had been a talented football player (dispelling the myth that all footballers are stupid*), and like most of the quantum-mechanics pioneers was a heavy drinker (Dirac and Ehrenfest were the exceptions). By 1933 he was generally regarded as the father figure of quantum mechanics, attracting funding to the field and arbitrating disputes, providing worldwide leadership. He was thus the perfect target for cynicism and sarcasm à la Ettore Majorana. A leader just had to be a senile old fart, according to Ettore.

But Ettore did warm up to Bohr as he grew to know him better, acknowledging his good nature, if not his intellect. He wrote to Gentile to describe how Bohr had invited him to his home, which, like the Copenhagen Institute, was paid for by

---

* He was for a while the goalkeeper of AB, the Danish champions on numerous occasions. His brother, a mathematician, was a prominent player for AB and made the Danish national and Olympic teams.

Carlsberg (as in "probably the best beer in the world"). The house was surrounded by such mountains of kegs that Ettore couldn't find the entrance—he needed Bohr's assistance to find the path through the beer labyrinth. Parenthetically, I note that much of the early research on quantum mechanics was sponsored by Carlsberg—a non-profit-making foundation. Spot how its bottles sneak into many historical photos. (One wonders if the most abstruse aspects of quantum mechanics, like the lack of determinism and the wave nature of matter, have deep roots in ethylization.)

Bohr's institute was the center of quantum-mechanics research in those days, and Ettore had a chance to interact with its luminaries. He met the very rude Pauli, but did not seem to notice the hiccups and complaints reported by others ("a *tipaccio*, very nice and intelligent" was his verdict). He met the Dutch physicist Paul Ehrenfest, who inquired at great length about his work, inviting him to Leiden later in the year. This visit would never take place. Ehrenfest—a very intimate friend of Einstein—suffered from severe depression, and in September 1933 shot his younger son (who suffered from Down's syndrome) before killing himself. But in the conversations between Ettore and Ehrenfest, there was no hint of dark thoughts from either side. The things that people think but don't mention.*

Later, Bohr left for a brief stint in the mountains with Heisenberg, where the two often retired to discuss philosophy. They were both deeply concerned with the interpretation of quantum theory. They spent their breaks trekking and mountaineering, all the while racking their brains up in the snowy summits, trying to understand the deep meaning of the new science. "When [Bohr] returns he will resume the apostolate to spread the Spirit of Copenhagen," Ettore wrote to Gentile on December 3, 1933. A hint of jealousy regarding the special relationship between Bohr and Heisenberg lingers in these words.

Ettore spent a short Easter break in Rome, going to Via Panisperna only a couple of times. He then returned to Leipzig, still in a good mood. On the train, he struck up a friendship with a drunk Neapolitan traveling to Germany to sell tomatoes.

---

* It's a curious coincidence that Ehrenfest's nickname in the quantum mechanics community was also the Inquisitor.

Ettore passed himself off as a fellow tomato seller, and the two became best mates for the duration of the trip. He arrived back in Leipzig in May of 1933, and for some time it was business as usual. And then something happened.

We simply don't know what, but it triggered the beginning of a long slump into depression. After his uncharacteristic openness in the previous months, his dark side returned with a vengeance. He resumed his habit of walking on his own and rarely talked to anyone. His extreme negative criticism became worse than ever, alienating those around him. He became poignantly antisocial. During a conference, Heisenberg praised Ettore's work and prompted him to intervene regarding a certain point of their theory; to everyone's embarrassment, Ettore responded with silence. He rarely mentioned Heisenberg now in his letters home, and when he did, his enthusiasm was gone. His tone became brusque, rude, blatantly meant to shake off human contact. He wanted to be alone.

Why this sudden break in his humor? We simply don't know, but he had followed a similar pattern before—for example, when he alternated gregarious happiness at Casina delle Rose with a more somber mood at Via Panisperna. But this time the swing was much more dramatic, and it would permanently damage his life. Are the roots of what happened in March 1938 to be found in May 1933? It's possible, but what exactly happened?

We can only speculate, but I'm sure the chronology can never be so simply linear. Whatever happened in May 1933 must have been in part the culmination of the events of the past two years. Ettore had endured a war of attrition on many fronts, steadily fraying his nerves. He'd had to contend with the specter of a charred baby, in an affair that came to a head in 1932. He'd been trying to evade the stifling Majorana family atmosphere and the overwhelming presence of his mother. The quarreling at Via Panisperna, with its "negative energy," had been getting worse: He'd been back there over Easter—we don't know what happened, but his tone to the Boys, Gentile excepted—immediately changed. Lack of love and of a woman can't have helped. And his health had finally given in to the bombardment of coffee and cigarettes. He may not have caught syphilis, as "suggested" by Segrè, but he did develop a bad ulcer while in Germany. Was this a cause of his troubles, or an effect?

Maybe it's pointless to seek out a first cause. Perhaps the point is that there *wasn't* one. Anyone can handle a single calamity; it's usually the combined effect of many blows that topples even the strongest. And the maelstrom whisking Ettore into depression had been steadily upping its beat. Given the numerous upsets in his life, could Ettore have been a victim of the proverbial last straw?

Regardless, we can identify one event that hit Ettore hard in May 1933. Ettore may have been insensitive to publication and career accolades, but that doesn't mean he didn't care about his science. And it was around this time that he suffered a nasty blow. For he was in Leipzig when news began to arrive that his critique of Dirac's theory, embodied in his 1932 masterpiece, was wrong.

THIRTEEN

His Unfinished Symphony

❀ ❀ ❀ ❀ ❀ ❀ ❀ ❀

To put it simply, the Majorana neutrino is what happens when Dirac's antiworld meets a shy and left-handed particle. But hiding behind Ettore's 1937 paper on the neutrino there's a complex human drama. By comparison, the scoops that haunted Fermi and Salam feel like prepubescent self-aggrandizement. Ettore's scientific tragedy, in contrast, had nothing to do with vanity. There's strong evidence that Ettore had finished his neutrino paper as early as 1932 and kept it in the drawer. He did this because the neutrino paper was nothing but a minor appendix to another paper he published in 1932, which he regarded as his real masterpiece. "Relativistic Theory of Elementary Particles with Arbitrary Spin," reads its title. I must prepare you for what follows. Seventy-five years on, I sometimes have the impression that physics is all wrong—and that the contents of that paper hold the key to our errors.

It's amazing that Ettore's 1932 masterpiece reached the public at all, given his history of abandoning his work. That he bothered to publish it suggests he knew his theory was so crazy it might even be true. But he published only in Italian; when he was in Leipzig he mentioned in a letter that he was working on an extension with a view to publishing in German, but he never finished it, or if he did, it was lost (much of his work went missing in strange circumstances). His masterpiece had to wait for a revival in the 1960s. It's a gem that was quickly forgotten and even today no one knows quite what to make of it. My personal view is that it

$$\cdots \left[\frac{1}{\cdots} + \frac{1}{\cdots}\right](a_j \cdots s+1) - \frac{2a_j}{s(s+1)} - \frac{2\cdot s+1}{(s+1)(s+2)}$$

$$\boxed{z_0 = 0}$$

$(s, m \mid a_x - i a_y \mid s, m+1) = \sqrt{(s+m+1)(s-m)}$

$(s, m \mid a_x + i a_y \mid s, m-1) = \sqrt{(s+m)(s-m+1)}$

$(s, m \mid a_z \mid s, m) = m$

$(s, m \mid b_x - i b_y \mid s+1, m+1) = -\tfrac{1}{2}\sqrt{(s+m+1)(s+m+2)}$

$(s, m \mid b_x - i b_y \mid s-1, m+1) = \tfrac{1}{2}\sqrt{(s-m)(s-m-1)}$

$(s, m \mid b_x + i b_y \mid s+1, m-1) = \tfrac{1}{2}\sqrt{(s-m+1)(s-m+2)}$

$(s, m \mid b_x + i b_y \mid s-1, m-1) = -\tfrac{1}{2}\sqrt{(s+m)(s+m-1)}$

$(s, m \mid b_z \mid s+1, m) = \tfrac{1}{2}\sqrt{(s+m+1)(s-m+1)}$

$(s, m \mid b_z \mid s-1, m) = \tfrac{1}{2}\sqrt{(s+m)(s-m)}$

$(s \mid b_z^2 x_0 \mid s+2) = c_{s+2} \tfrac{1}{2}\sqrt{(s+m+1)(s-m+1)} \cdot \tfrac{1}{2}\sqrt{(s-m)(s-m+2)}$

$(s \mid b_z x_0 b_z \mid s+2) = c_{s+1} \cdot \tfrac{1}{2}\sqrt{\quad''\quad} \cdot \tfrac{1}{2}\sqrt{\quad''\quad}$

$(s \mid x_0 b_z^2 \mid s+2) = c_s \cdot \tfrac{1}{2}\sqrt{\quad''\quad} \cdot \tfrac{1}{2}\sqrt{\quad''\quad}$

$(s \mid b_z^2 x_0 \mid s) = \tfrac{1}{2}(s+m+1)(s-m+1) c_s + \tfrac{1}{2}(s+m)(s-m) c_s$

$(s \mid b_z x_0 b_z \mid s) = \tfrac{1}{2}(s+m+1)(s-m+1) c_{s+1} + \tfrac{1}{2}(s+m)(s-m) c_{s-1}$

$(s \mid x_0 b_z^2 \mid s) = \tfrac{1}{4}(s+m+1)(s-m+1) c_s + \tfrac{1}{4}(s+m)(s-m) c_s$

• $c_s = \tfrac{1}{2}(s+m+1)(s-m+1)\,\Delta c - \tfrac{1}{2}(s+m)(s-m)\,\Delta c$

$= \frac{2s+1}{2}\,\Delta c$ $\qquad \Delta c = c_{s+1} - c_s = c_s - c_{s-1} = \Delta$ [3.40]

$c_s = (s+\tfrac{1}{2})\Delta$

carica della particella −e

$$\boxed{\left[a_0\left(\frac{W}{c} + \frac{e}{c}\varphi\right) + (a, p + \frac{e}{c}C) + mc\right] \psi = 0}$$

$\beta = 1$

$b_m = \tfrac{1}{2} e_3 \, \epsilon_x$

$a_0 = e_1, \quad a_3 = e_2 \epsilon_x$

$i(b_x, a_0) = \tfrac{1}{2}\epsilon_x (e_3, e_1)$

$= -i e_2 \epsilon_x$

$a_x = i(b_x, a_0)$

$a_y = i(b_y, a_0)$

$a_z = i(b_z, a_0)$

della $a_0 = s + \tfrac{1}{2}$

$(s, m \mid a_x - i a_y \mid s+1, m+1) = -\tfrac{i}{2}\sqrt{(s+m+1)(s+m+2)}$

$(s, m \mid a_x - i a_y \mid s-1, m+1) = -\tfrac{i}{2}\sqrt{(s-m)(s-m-1)}$

$(s, m \mid a_x + i a_y \mid s+1, m-1) = \tfrac{i}{2}\sqrt{(s-m+1)(s-m+2)}$

$(s, m \mid a_x + i a_y \mid s-1, m-1) = \tfrac{i}{2}\sqrt{(s+m)(s+m-1)}$

$(s, m \mid a_z \mid s+1, m) = \tfrac{i}{2}\sqrt{(s+m+1)(s-m+1)}$

$(s, m \mid a_z \mid s-1, m) = -\tfrac{i}{2}\sqrt{(s+m)(s-m)}$

A page from Ettore's notebook, where he developed his infinite component equation, meant to replace Dirac's equation and do away with the ungainly negative energies.

set up a fork on the road of twentieth-century physics, and we didn't take Ettore's path. We may yet live to regret it.

Ettore didn't like Dirac's theory. A sea of infinite particles with negative energies being defined as the vacuum? Nah, it sounded bogus. His original motivation in writing his masterpiece was to do away with Dirac's negative-energy sea and the prediction of antimatter. Ettore examined the root of the problem with a clinical eye, deconstructing Dirac's elegant mathematics and homing in on the source of the negative energies, and he decided to do something entirely orthogonal. It's hard to explain without mathematics, but it's beautiful, even more so than what Dirac had done. Mathematical beauty isn't easily explained. As Fernando Pessoa stated in a poem: "Newton's binomial is as beautiful as the Venus of Milo; it's just that very few people notice it."

Essentially, he allowed himself to be guided singly by the principle of relativity—that there are no preferred observers in the universe—and didn't require his equation to describe an electron from the start but to describe anything with spin and consistent with relativity. Only at the end did he require the electron to be an aspect of this "anything." More mathematically he considered infinite dimensional representations of the Lorentz group. In fact, what Ettore did is a more obvious way to unify relativity and quantum mechanics than Dirac's theory. I find it a bit surprising that Dirac in 1927 didn't follow Ettore's path.

Dirac had been shocked to find two types of particles emerging from his equation: matter and antimatter. Ettore found instead an infinite tower of particles, all described by the same equation, unified into a single quantum-mechanical wave. At low energies they seemingly split into distinct particles (including the electron), unfolding a particle rainbow, each "color" with a different spin and mass. But as with Dirac's electron and positron, at a deeper level they were in fact the same object showing different faces. The crucial fact that the various particles in Ettore's tower had different masses (unlike the electron and positron) allowed him to ignore all but one (and to "decouple" the rest, to use the modern terminology). He didn't need to appeal to a Dirac sea because there were no negative energies or antiparticles in his theory.

Ettore was very proud of his accomplishment. His construction was even more graceful than Dirac's; by Dirac's own aesthetical token, it should therefore have been correct. In Leipzig, Ettore continued to polish his masterpiece while he worked on strong nuclear forces with Heisenberg. He referred to it in his official

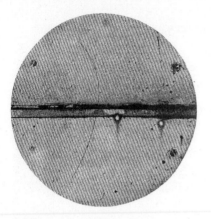

The innocent track seen in this picture led to the discovery of the positron and the antiworld.

reports sent to Italy and in letters to his father and to his friend Gentile. It's obvious he valued it more than anything else he'd done. Not many understood it, but he didn't care. He knew Dirac's theory was tripe, both theoretically and experimentally. Just to mention one problem, where was Dirac's so-called antielectron or positron, anyway?

It was while Ettore was in Copenhagen, in March 1933, that news started to trickle in that the positron had been discovered by American physicist Carl Anderson. Examining cosmic rays in a cloud chamber subject to a magnetic field, Anderson had seen the telltale curling track of a particle with a mass and charge like the electron, but bending in the opposite direction, implying that it carried a charge with the opposite sign. At first Anderson couldn't believe his eyes. Then he let out a sigh of relief. *Of course* his prank-loving friends had just reversed the polarity of his magnets. The idiots! But when he saw that no one had tampered with the magnets, it gradually dawned on him that he'd discovered something new. A bit shaken, Anderson eventually published a cautious note titled "The Apparent Existence of Easily Deflectable Positives." Only much later did he make the connection with Dirac's positron.*

No one believed Anderson's results. Pauli, who, like Ettore, hated Dirac's theory, poured scorn all over Anderson. Even Dirac, who had a vested interest in these findings, was skeptical. Niels Bohr went as far as to suggest that Anderson's anomalous tracks were caused by air currents in the cloud chamber. "Here everyone has taken it for a formidable equivoque," Ettore reported sarcastically in a letter from Copenhagen. "The positive electrons are nothing but normal electrons [moving in the opposite direction]. This is also Rutherford's opinion."

---

* Anderson coined the term *positron*, a contraction of *positive* and *electron*; but he also wanted the electron to be renamed *negatron* for consistency. There were other proposals: The eccentric Herbert Dingle, recalling that Electra's brother is called Orestes in Greek mythology, suggested that the antielectron be named Oreston.

All this cheerfulness on the matter dissipated after the Easter break. In England, Patrick Blackett and Giuseppe Occhialini independently confirmed Anderson's discovery. Pauli was so incensed by this little twist of nature that he went on holiday to southern France, so he wouldn't have to hear news of the confirmation of the positron. But the evidence in favor of the positron kept mounting, and even Heisenberg began working on Dirac's theory. "Season's news," Ettore wrote in a letter, "credence is being given to Dirac's theory of the positive electrons. Heisenberg is working on it seriously." Ettore then makes fun of the prediction of creation and annihilation of pairs of matter and antimatter particles in Dirac's theory, but the joke falls flat. Soon afterwards Occhialini and Blackett observed precisely these outlandish creations and annihilations. Dirac was right; and by the most linear extension, Ettore's theory was dead.

Ettore lost heart. I don't know why: He'd only proposed a theoretical idea that wasn't upheld by nature. So what? That's science. Better to have a good idea than a bad one or none at all, even if nature doesn't choose our idea. But he took it very personally; some suggest that this was the trigger for his 1933 fall into depression.

I'm not sure how Dirac would have reacted if the positron had never been found and instead Ettore's infinite tower had been discovered in cosmic rays. He probably would have fared better than Ettore: Even with all of Dirac's emotional scars, Ettore's mind was on much shakier ground. But seventy-five years later, the story is far from finished. Ettore's 1932 paper is a box of surprises, quite apart from the fact that it led to an amazing prediction about the neutrino. As Dirac himself put it, "The opposite of a correct statement is a false statement; but the opposite of a profound truth may well be another profound truth." Ettore's and Dirac's mathematical constructions are so beautiful that being opposites doesn't preclude them from being complementary, and possibly both true.

The problem with Ettore's paper is that it was too far ahead of its time. I have a sneaking suspicion that it may be the most lasting contribution he left us. *Ettore's theory is a gem in unification*. And we badly need it.

Current particle physics is built upon the so-called standard model, which is a glorious mess. Twelve fundamental particles make up matter, another four carry forces and mediate interactions between things, one "auxiliary" particle is thrown

Three Generations of Matter (Fermions)

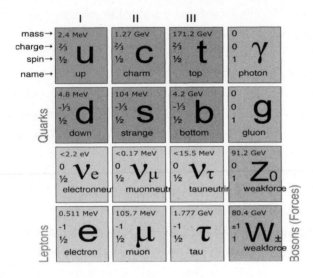

Figure 13.1: The standard model of particle physics. There are enough so-called "fundamental" particles to compose a periodic table.

in for good measure, and this still neglects gravity (see Figure 13.1). The electron and neutrino have been joined by two more families of similar but more massive particles, the muon and the tauon, plus their corresponding neutrinos, the whole lot called leptons. The proton and neutron are no longer fundamental, but are instead known to be made up of two types of "quark" (the up and down quarks.) As if this weren't enough, four more quarks have been discovered (the charm, strange, top, and bottom quarks). Fermi's weak force is known to be carried by the W and Z bosons, while light and electromagnetic interactions are transmitted by photons. Ettore and Heisenberg's strong force is mediated by gluons. We need the Higgs particle to give mass to particles, if appropriate.

A hell of a zoo of building blocks for the world! (To think that the Greeks were happy with earth, air, water, and fire.) Anderson may have received the Nobel Prize for discovering the positron, but nowadays the view is that the finder of a new elementary particle ought to be punished by a $10,000 fine. And if you consider all the "free parameters" that have to be dialed to exactly the right values for the model to agree with the data, then between masses, interaction strengths, mixing angles, and so on, you have almost thirty independent numbers. All thrown into the theory by hand, as input, without a glimmer of justification.

Pathetic.

So far, attempts to improve the overcomplicated state of particle physics have focused on unifying forces; i.e., turning the many particles mediating interactions into a single unifying force particle. For example, Fermi's weak force, electricity, and magnetism now fall under the single umbrella of the "electroweak model," discovered by Salam, Steve Weinberg, and Sheldon Glashow. Some voice hopes that the strong force (or even gravity) may join the "interactions" party. All very well, but this still leaves us with a ridiculous proliferation of fundamental particles that don't carry forces: quarks, leptons and the Higgs: the forgotten side of uni-fication. To my mind, this signifies one of two things. Either these particles are composites—little "molecules" composed of smaller and simpler constituents we haven't yet discovered; or we should seriously revisit Ettore's 1932 theory.

The underlying idea is easy to grasp: No one ever counts electrons and positrons in Dirac's theory as two different types of particle. They're just two different states of the same particle, much like spin up and spin down for a spin half particle. A sin-gle quantum wave describes the doublet: They're different "entries" of the same object. By analogy Ettore's construction throws us a lifeline in unifying the whole wild menagerie of particles we insist on calling fundamental. Perhaps they are really manifold expressions of the same fundamental object, just like the electron and the positron. The different states in Ettore's infinite tower have different masses and spins, and these don't match experiment. His symphony is unfinished. But an *adap-tation* of Ettore's 1932 idea just might provide the ultimate unification.

After his 1933 descent into depression, Ettore was quoted by Amaldi as saying, "Physics is on the wrong road, we're all on the wrong road." This quote has made its way into Italian comics, with references to atomic weapons and visions of Ar-mageddon. But it could just as well be applied to the state of modern physics. We had a choice in 1932, and we took Dirac's path, when in fact we never actually needed to choose one path exclusively (it's possible to fold antiparticles into Et-tore's scheme). We got ourselves into a terrible muddle and are now appealing to string theory, supersymmetry, and other complex constructions containing infinite towers of particles that don't fit the observed world any better than Ettore's infinite dimensional wave function. Ettore, with a simpler approach, in a sense had already achieved as much as string theory—it's ironic that what excites some people so much nowadays is no better than what for Ettore seemed an utter failure. But why not go back to where he left off and *adapt* his construction to clear up the chaos pervading our cherished "standard model"?

"In all probability Majorana knew exactly what had happened . . . and had guessed the destructive potential of the discovery. . . ." Ettore's speech balloon: "Physics is on the wrong road . . . we are all on the wrong road . . ."

Late at night, I often have a lingering feeling that a variation on Ettore's theory may provide the essential clue. It could unify in a single particle what then unfolds into the fundamental-particle avalanche we're struggling with. But his 1932 masterpiece is not only unfinished: It has largely been forgotten. As for its appendix, left to gather dust in his drawer for almost five years, that certainly can't be ignored. I know I advocated a Nobel Prize for Ettore based on the Majorana neutrino, but this work was insignificant in comparison to the majestic views he unveiled in 1932. The Majorana neutrino was a mere sidekick to a forgotten, ill-understood masterpiece: Majorana's unfinished symphony.

It's always tough to be disproved, but was this the only cause of Ettore's sudden collapse in 1933? One certainly cannot discount the effects of the main undercurrent to all the waves in these stormy seas: the specter of the burnt baby—the leitmotif of his tragedy. It was a curse that had been continuously hacking at his soul for almost a decade, sapping precious stability and coherence, year after year wearing away the remainder of his certainties. The charred-baby blight changed Ettore's worldview forever, and had a similar effect upon his brother Salvatore. Neither was ever the same again. And when its troubles finally relented, in the months just before his visit to Leipzig, it may well have been too late for Ettore to feel any relief. Permanent soul damage had been sustained.

FOURTEEN

The Hand That Rocks the Cradle

✾ ✾ ✾ ✾ ✾ ✾ ✾ ✾

The story of the burnt baby—"the outrageous Majorana affair"—is so improbable that it may prove useful to consider a fictional prelude: a short story from *La Paura di Montalbano* by Andrea Camilleri; a tale well tuned to the world of Sicilian vendetta.

Detective Salvo Montalbano, a multibook character renowned for his cynical attitude toward Sicilian officialdom, comes across a story that has its roots more than fifty years in the past. An old lady on her deathbed confesses to a priest, who is, naturally, bound to silence. But the priest then drops disturbing hints to the detective.

The old lady had had a friend who'd asked her for rat poison many years before, while going through a difficult phase with her husband: a good-for-nothing who slept around, spent his nights playing cards, and ran up large debts that his wife then had to pay. The old lady had easy access to poison, since her own husband, recently deceased, had left her a pharmacy. But seeing the obvious intent in her friend's request, she gave her a harmless white powder instead.

A few weeks later, her friend's husband felt unwell while playing cards late at night. He was brought home, the doctor reporting that it was nothing more serious than the effects of excess. He recommended rest, less drinking and smoking, and left the wife a medicine—a white powder—asking her to administer it to the patient. The next morning, the wife was still locked in their bedroom when the servants broke in to find her in a state of shock, her husband bathing in his own shit and vomiting his guts out.

He died shortly afterwards, and the wife promptly confessed that instead of giving him the medicine, she'd poisoned her husband. Fifty years later, the old lady who'd provided the "poison" still feels guilty and insistently murmurs to the priest, "It was *not* rat poison!" before she expires.

Given the long period of time that has elapsed since the original events, it takes Montalbano quite an effort to find out what actually entered the official records. It appears that the wife did go to prison at first—after all, she confessed her crime. However, the case was reopened several times. Her father was an important politician (read: a powerful mafioso) who was engaged in open warfare with several political enemies (read: competing Mafia factions). He claimed that his daughter had been forced to confess under duress by his enemies and ordered an autopsy in Palermo. After careful testing, the doctors found no evidence of poison in the husband's corpse. The woman was promptly released . . . still claiming that she really had poisoned her husband.

Naturally, her father's enemies argued that he had bribed the doctors and asked for an independent autopsy to be carried out in Florence. The body parts, however, were lost on the way there, and when they eventually arrived six months later, had quite possibly been tampered with. When the second examination was finally performed, it did reveal poison . . . and in such large amounts! Questioned as to how the Palermo doctors could possibly have missed it, the Florence doctors explained that they mustn't have tried "the right chemical." The poison they'd found could only be detected by means of one particular chemical. The woman returned to jail, despite her father's protests.

The archetypal Sicilian mêlée is now installed: what a richness of layers, what depths of corruption! You can see Montalbano's dilemma: A woman seeks revenge on her husband with poison that isn't poison; however, her husband does die from poisoning. Palermo doctors, paid by her father, find no poison in the body, whereas an "independent" inquiry committee, bribed by his political enemies, does find poison in the body. There are at least two levels of vendetta, but could there be yet a third? Could someone else have poisoned the husband? But then what a coincidence. . . . This is genuine Sicilian intrigue, strand after strand of political and emotional battle intertwining, different layers of feuds and revenges combining in a web where cause and effect can no longer be distinguished.

But amid the confusion, the clever detective smells a rat. Why did the old lady feel so guilty about this incident? How could she *not* be sure she'd given her friend an innocuous substance? It's not a mistake easily made. And if she did know that it wasn't rat poison, it would surely have been enough to tell the police her version of the story to get her friend's sentence reduced, the crime diminished to mere *intention* to murder. Was she protecting someone? What else might be hiding beneath the topsoil?

It is then that two developments come to the aid of the detective. First, he finds out that while the Palermo doctors were writing their report, one of them had died. Another doctor had stepped in to complete the work, and it appeared that after backhanding the initial doctors the mafioso father *hadn't* bribed the new doctor. Being scrupulous, the new doctor had repeated all the tests himself before signing the report. And although this detail was included only in his case notes—not the report itself—he *had* tried the missing chemical alluded to by the Florence doctors. With a negative result.

To Montalbano this proved beyond doubt that the man hadn't been poisoned, but had died of natural causes. Even though both sets of doctors were corrupt, the Palermo doctors didn't need to lie, so the final Palermo doctor didn't even have to be dishonest.

Second, he learns that the old lady left Italy to live abroad shortly after the original "poisoning" incident. She sold everything, including the pharmacy left to her by her husband, and Montalbano discovers that the new owner is still alive and is the sort of old man who likes to talk at length about the past. The old pharmacist recalls that one day, while the woman was checking the business records before completing the sale, he'd caught her in great distemper, indeed shouting with rage, her nerves completely out of control. He'd never found out the cause but the old man manages to place this episode just before the poisoning occurred; i.e., *after* the old woman had given the fake poison to her friend. All then becomes clear to Montalbano.

He only needs *proof*, and illegally breaking into the place where the old woman's belongings are stored confirms his theory. Among her possessions he finds several letters between her late husband and the "friend" to whom she'd given the powder. Her husband and friend had been lovers and the letters are cruel, unforgivable. Not only were they cheating on her, but they made fun of her in their correspondence: of her sexual frigidity, her character, her physical flaws.

The old woman felt guilty on her deathbed not because she might have given her friend rat poison by mistake. Her silence, regarding the fact she *knew* it was *not* rat poison, was the third and final level of vendetta, and the cause of her guilt. She'd been responsible for her friend's long jail sentence, and worse, for her friend's lifelong remorse for having "killed" her husband. The priest enlisted Montalbano because the friend turned out to be still alive, presumably still believing she'd poisoned her husband. But Montalbano, having satisfied his curiosity, does precisely nothing about it.

Contrived? You read worse things in *La Sicilia* newspaper. Or closer to the topic of this book, in the history of the Majorana family. Except that the burnt-baby case is only formally similar to Montalbano's story: Its contents are infinitely more gruesome and the emotions involved much more disturbing. Not to mention that the Majorana "case" was enacted by very real people—just ask the innocent baby who was caught in the middle.

The dead-baby case was closed shortly before Ettore's collapse, but its inception goes back some eight years, to the summer of 1924, when Ettore was eighteen. At the house of Antonio Amato, a wealthy confectionary industrialist, a terrible fire broke out in the bedroom where his only son—a baby—was asleep. Tongues of fire soon engulfed the cradle, turning the space between mattress and mosquito net into an inferno and burning little Cicciuzzo Amato to a crisp. Although the incident was initially seen as a misfortune, closer examination revealed that the cot had been doused with a flammable liquid. Further police investigation, Sicilian style, extracted a confession from the culprit: a sixteen-year-old nursemaid, Carmela Gagliardi, who turned out to be mentally retarded.

In the logic of the half-wit, she explained to the police that her family had forced her to work for the Amatos whereas she wanted to work for another family, the Platanias, who were fond of her and to whom she had become attached. Therefore she had burnt the Amatos' baby.

Sic.

Besides being simple, the poor girl had good reasons to be emotionally disturbed. Her brother sexually abused her; her mother spanked her on a regular basis; she had to work hard while her sister idled away at home; the same sister was betrothed to a boy Carmela had dated and still loved and who had previously sworn eternal love to her before assaulting her . . . the ubiquitous Sicilian cataclysm.

But just as in Montalbano's chronicle, or any other good Sicilian tale, one layer of vendetta doesn't suffice; and another was already running in parallel, ready to pounce upon the case. It concerned the Majorana family. As chance would have it, two of Antonio Amato's sisters were married to Ettore's uncles Giuseppe (the eldest brother) and Dante (the fourth). When I described Ettore's family, I hope I gave you the correct impression: These were men of great distinction—jurists, deputies, senators, and rectors.

For reasons not clear to me, Antonio's sisters had been cut out of the family will, which Antonio had thus collected whole. But in Italy the law compels a certain portion of every will to be given to *all* the children, regardless of the wishes of the testator.* *"La legitima"*—the legitimate, or the fair one—prevents any child from being entirely cut off and left destitute. Antonio had chosen to ignore this law, but Dante and Giuseppe, as knowledgeable lawyers, requested that he restore *la legitima* to his sisters. A battle then developed along the following improving lines:

The Majoranas proposed that the matter be amicably settled by means of sum X. Antonio countered by offering one-fifth of X. Counteroffers were made. Antonio didn't budge. The matter was taken to court, the Majoranas' natural environment. The court ruled that Antonio should pay seven-fifths of X; i.e., 40 percent more than the Majoranas originally asked for. Antonio was left with a big chip on his shoulder.

I leave it to the reader to work out who started "hinting" to the police that the reasons given by Carmela Gagliardi, the retarded maid, weren't good enough; that someone else must have been behind her crime, and that Dante and perhaps Giuseppe should be investigated. Suitably buttered up, the police put "pressure" on Carmela, who duly produced a second confession.

In true soap-opera style, however, the plan backfired. Instead of incriminating the Majoranas, the girl, perhaps predictably, chose to take revenge on more immediate targets who had slighted her. First of all, her previous lover, now betrothed to her sister: It was he, she said, who'd given her a white glass bottle filled with

---

* Specifically, you can only decide where half of your estate goes; the other half has to be equally divided between all surviving direct relatives.

petrol to burn the baby. Then her mother and brother: They were the ones, she insisted, who had forced her to do the terrible deed against her wishes. For some reason she spared her sister; maybe leaving her out of jail was part of her sweet revenge. But with four people now involved in the affair, it took several twists, turns and further twists to bring the blame back to the Majoranas, as initially intended. At any given time at least one of those imprisoned denied any involvement.

A few years later, after much bribery, all jailed parties finally agreed on a story. By then, Carmela's brother had gone mad, which didn't deter the police. The case was reopened. Yes, they were all part of a conspiracy to burn Amato's baby and they had acted at the command of Dante Majorana. He'd given them money as well as a green bottle filled with benzene. The fact that Dante had no motive to take revenge on Antonio (quite the reverse) was ignored in court. And how the white petrol bottle changed color and content in the intervening years also seems to have gone unnoticed. Dante Majorana was jailed, together with his wife Sara, sister of Antonio. The second level of vendetta was complete.

It then took three long years for the defense to dismantle the prosecution's case. Part of the problem was that the judge was compelled by law to follow the rule that "children always tell the truth" (!?), and therefore Carmela Gagliardi couldn't have lied. Furthermore, Italian law at the time ruled that a second confession is more reliable than the first. Against these absurd rules, the defense produced evidence that witnesses had been roughed up by the Mafia, that the Majoranas lacked any motive, and that the forensic proofs produced by the prosecution were deeply flawed. But I suspect that, in the end, the Majoranas may simply have resorted to counterbribery.

Eight years after the case had started, the by now grown-up Carmela, sobbing and sniveling, confessed for the third time: "I alone am guilty." I'm amazed this stood up in court, given that she was no longer a child. But as Sciascia puts it, "Were it not for her tears, her remorse, no one would have remembered that, at the heart of that labyrinth of hate, falsehood, and despair lay the little corpse of Cicciuzzo Amato, the baby burnt to death in his cot."

The murdered-baby tribulations affected Ettore tremendously. Laura Fermi wrote that "Ettore wanted to prove his uncle's innocence, clear him of a suspicion that could not fail to taint the entire Majorana family. . . . The ordeal was too great a strain on Ettore's sensitive temperament. After his return from Germany Ettore became a recluse." Amaldi, who kept in touch with Ettore during his dark period of reclusion, stated that "This incident would have had a decisive influence on Ettore's attitude to life." The family has protested against these views. Ettore wasn't a blood relative of the burnt baby. But he was very close to the accused uncle and aunt.

It's been said that he wrote to his uncle Dante "almost every day" while he was in jail. Another source said that he was so incensed at the court's hypocrisy that he offered to defend his uncle in court, putting his logical skills into action. "I don't believe in lawyers," Ettore told physicist Gleb Wataghin. "They're all idiots; I shall write the defense of my uncle myself: I know what happened, I've talked to him."* I very much doubt whether anyone took Ettore's opinions seriously: After all, the court case was not a matter for logical argument but for political manipulation and kickbacks. Ettore meant well, though, and this incident reveals much about his personality. In an ideal world things *would* be resolved in the way Ettore proposed. As it was, mathematical logic had nothing to do with the court trial; indeed, justice had nothing much to do with it either.

It certainly must have disturbed a person as private as Ettore to see his name associated with what was referred to in the sensational press as the *"delitto della culla"* (the cot murder) or the "outrageous Majorana affair." And then there is what might lie beneath the surface.

Let's face it: None of this makes any sense! It just doesn't square, as was the case in Camilleri's story. Burning a baby is an act of such extreme violence—even for an emotionally and sexually abused mental retard—that I have to believe yet another layer must have been present, hidden below the two surface layers, just as in Montalbano's tale. A cornuto tier, perhaps: a crime of passion of some sort, motivated by stronger venom and malice.

I don't want to start a conspiracy theory, let alone suggest anything concrete. But I'm absolutely sure there must be more in the "outrageous Majorana affair" than meets the eye, providing a final sense of logic for that which, on the surface, has none. Perhaps this hidden layer is even more "outrageous" than the obvious two—perhaps it has the simplicity of the one uncovered by Montalbano. And who knows how it may have mingled with Ettore's own tragedy; or not.

If only we had a Montalbano in attendance to unravel the mystery.

---

* Interview given in 1975 at the University of Campinas, São Paulo; Wataghin doesn't date this incident, though.

FIFTEEN

Stellar Collapse

✣ ✣ ✣ ✣ ✣ ✣ ✣ ✣

The burnt-baby incident permanently injured Ettore and his brother Salvatore. They were both naïve intellectuals who had led very sheltered lives. As the story unfolded and became more sordid, it shook the foundations of their worldviews. That the affair was eventually solved by the Majoranas by descending to the same level as their detractors (and by the men of the world in the family, rather than either of them) only made matters worse. The family publicly cleared its name, but in their eyes only dirtied itself even more. It did not remove their need for exculpation.

It's been said that Ettore lost faith in science; that he became sensitive to the alleged contradictions between science and religion; that his tragedy reflected that of Pascal, the French philosopher who rejected science for religion. But Ettore may have lost faith in more than just science. He may have lost faith in rationality altogether. In 1933 he descended into madness for numerous reasons, but it is indisputable that the baby incident contributed a crucial spice to the curry. We're all shocked by the story; but for Ettore and Salvatore—ivory-tower bookworms who believed in logic as the universal weapon—the burnt-baby affair symbolized the breakdown of rationalism. And neither fully recovered. Sending all to hell may, then, have felt like an adequate response: "Well, if you're going to be that base, then take this, motherfuckers." And they retreated into their shells. In Ettore's case it only took writing a letter.

֍

Here's what Ettore did to the Boys that was "downright inexcusable," thus ensuring they'd leave him alone. On May 22, 1933, he sent a certain type of letter to Emilio Segrè—the Basilisk—knowing damn well how touchy and insecure he was. This letter would be the subject of much public scrutiny many years later. An Italian newspaper titled an article about it: "The Surprising Letter Revealed by Nobel Laureate Segrè: Majorana Liked Hitler." Yet for anyone who examined the letter even superficially, it would be obvious that its point wasn't political. It was a personal assault. This letter would cut Ettore off from Fermi and the Panisperna Boys; that was conspicuously its intended purpose.

"Caro Segrè," begins Ettore, before adding a couple of formulaic greetings. He then complains about Dirac's theory becoming the flavor of the day in Leipzig and the world. He reports on the apparent tranquility of the political situation in Germany, quickly moving on to the core of the letter: a lengthy discussion of the "Judaic question" in Germany. It's May of 1933.

Five long, dry paragraphs begin with "The question of anti-Semitism [here] should be seen in the context of the [Nazi] revolution which eliminated wherever it could all opponents, among which were almost without exception the Jews. This is not to say that there isn't in Germany a grave Judaic question in itself and by itself. . . . The Hebraic question in Germany is quite different from that in Italy, be it by the spirit or number of the local Jews."

The tone is objective and emotionless, and if I didn't know the context I'd say it was uncharacteristically dull, considering Ettore's lively writing style. He quotes a profusion of statistics explaining how German Jews had come to dominate certain areas of public life; he explains the extent to which this led to social resentment. He states the facts underlying racism and the persecution of German Jews without comment. He never expresses an opinion. He details the philosophy of racism developing in Germany with minimal use of adjectives. Paragraph after paragraph drones on without much color or emotion unless one can read between the lines.

With one dramatic exception. Halfway through his letter, out of the blue and against the grain, Ettore announces that "it's been stated that the Judaic question wouldn't exist if the Jews were acquainted with the art of keeping their mouths shut."

Now, Segrè was Jewish, and he most certainly hadn't mastered that helpful art. We know about this letter because Segrè had it published in 1988, just before he died. He could hardly conceal how much it irritated him: "It's strange that [Majorana] wrote this letter to me, who certainly didn't appreciate it. . . . I like to think that if Ettore Majorana had lived longer, he'd have seen things differently and would have repudiated this letter."

The views Ettore expressed are undoubtedly indefensible: There is an undertow of tacit approval of German anti-Semitism throughout. But it's important to put this letter in its historical context, even before examining the personal background. This is 1933, and expressions already in use regarding the German Jews—such as "surgical intervention" or "final solution"—didn't yet have the meaning they'd acquire after the holocaust.* Ettore's letter to Segrè also contradicts the rest of his correspondence. A week earlier, in a letter to his mother, he stresses the disingenuous opportunism of the Aryan majority in their dealings with the Jews. In a letter to Gentile, dated June 6, 1933, he ridicules German racism: "Germany, not finding in her culture and history sufficient elements to establish a unifying feeling among German speaking peoples, is constrained to appeal to the silly ideology of race, which seemingly doesn't find an echo in Austria. Also the anti-Semitic fight, partly justified by instinct, isn't well supported by the arguments usually invoked, among which sadly dominates the eternal theme of race, and it's likely that it will die down quickly."

In other words, Ettore plays anti-German to his Fascist friend Gentile, and pro-German to his Jewish correspondent. If it weren't for the other events taking place in his life, we might think this was an intellectual game of "devil's advocate." But in the context of "retreating into a shell" and "leaving the world," it's obvious how to interpret Ettore's letter: It's anti-Segrè rather than anti-Semitic. And it would be almost funny, all these years on, were it not for the fact that Segrè lost his whole family in the holocaust.

---

* Segrè himself stated in his book on Fermi that the German anti-Jewish crimes at this stage were "minor."

Ettore's letter achieved its intended effect—he was duly severed from further interactions with the Boys. On his last official report, composed just before he returned from his working spell in Germany, he provided his home address for correspondence, not the Via Panisperna Institute. But *why* did he want to break with the Boys?

It's true that he was bypassed for promotion (a fact he must have learned around May 1933). After Heisenberg's blessing and his 1932 masterpiece (despite its flaws, real or perceived), it's possible that Ettore expected recognition—an official position, perhaps. He didn't get it. But he'd never shown any interest in career academia and so wasn't really slighted. It's also true that in 1933 Fermi and Amaldi were happily publishing Ettore's 1928 work without even acknowledging him. But Ettore had never cared about such things.

The fact is that he wanted to break with the world, not just with the Boys. His mood towards his family wasn't any better. When his family drove to visit him in Germany in July 1933, he dispatched them rather forcefully, refusing to join their European tour. Later he shook off his mother's insistence that he rejoin them in the resort of Abbazia in August. And when she threatened to travel to Rome to be at home when he returned from Leipzig, he replied in a letter, "Your concerns with my intention to go directly to Rome seem to me exaggerated. . . . You'd give me unnecessary displeasure by undertaking such a long trip without a good reason or purpose. But I don't intend to change my plans for fear that you'll put into action such an irrational threat."

On a hot Sunday in August 1933, Ettore returned to Rome from Leipzig. He found the enormous house in Viale Regina Margherita empty. He was completely by himself, as he wished. And thus began his silent years. From that day until 1937, he barely left his bedroom. Even after his family returned, he'd take in food—usually just milk—leave out a chamber pot, and keep ablutions to a minimum. He shared his room with Luciano, but his brother was hardly ever at home during this period. When Luciano was home, he saw Ettore always at his desk, working away. Whatever he was doing has since been lost.

Despite his strange behavior, the family let him be. He grew his hair and beard to proportions that wouldn't be socially acceptable until well into the 1960s. His friends sometimes sent a barber to tame his hippie looks despite his protestations. Those who visited him said that he acted like a terrified man. But he didn't tell them much: His heart became intangible.

Amaldi, the most self-effacing of the Mark I Boys, tried to keep in touch. Against Ettore's wishes, he visited on a regular basis. He reported on the developments taking place at Via Panisperna: how Fermi had finally come to believe in the "neutral proton," as Ettore had called it; how he was now becoming a world authority in the matter; how they were experimenting with slow neutrons. This is allegedly when Ettore uttered his infamous statement: "Physics is on the wrong path; we are all on the wrong path." A chilling statement given that the Boys were about to bombard uranium with slow neutrons, not realizing what they were doing.

Ettore's only expressions of warmth were reserved for Giovanni Gentile, who sent him a book with the dedication, "To a dear friend, of whom I no longer know anything." Ettore wrote back, with a pale shade of his earlier sarcasm. Over the next four years, Gentile continued to send books. Ettore always acknowledged them, albeit in letters that became briefer and briefer. He also kept up a skeletal correspondence with his uncle Quirino.

Ettore's overall behavior has been catalogued as a "nervous breakdown." It wasn't: He could go in and out of this "mad" state at will, as is well attested by the surviving documents of the period. It is also clear that while he carried on with his science, his interests diversified. He became intrigued by medicine (the physiology of the brain) and by what would now be called game theory (he used real navy data in mathematical war games, lending fuel to conspiracy theorists). He wrote a paper on sociology, which Gentile had published after Ettore's disappearance. He worked hard on theology, rejecting the "vulgar materialism of science." He became an expert on the pessimist philosopher Arthur Schopenhauer.

Strangely, in 1935 and 1936, Ettore even proposed to give a course at Via Panisperna. We know this because the paperwork has survived. Page 1 contained the syllabus (his 1932 article plus the second quantization of the electromagnetic field). Page 2 was meant to be filled out by Corbino, assigning credits to the course and making a lecture room available. But Corbino never followed through. Was Corbino's lack of action malicious? Either way, Ettore's proposed course would definitely have been too hard for the students.

The lines of the "obscure spiral" were now combined: a failed theory, a disastrous personal and family life, wrecked health, alienation from the scientific community. . . . Do we need more?

Just to make matters worse, in 1934 Ettore's father died, hammering the final nail into his coffin of despondency. How badly did he suffer from this? In my research I often found that my interviewees used Ettore as a mirror onto themselves. But in this particular case I found someone who may instead mirror Ettore's soul. The self-styled "last wheel of the Majorana cart": Luciano's third son Pietro, Fabio and Ettore Jr.'s younger brother. His father died before Pietro had a chance to get to know him properly. And it's precisely in these waters that I found a reflection of Ettore's soul.

SIXTEEN

Artichokes

✢ ✢ ✢ ✢ ✢ ✢ ✢ ✢

On the Italian Riviera, near the French border—with the splendors of Monte Carlo not far off—is a shabby seaside resort that has seen better days. Once the playground of the Russian aristocracy and Alfred Nobel's last home, Sanremo is now replete with modern bad taste mixed up with faded grandeur: a dodgy casino, a profusion of Russian prostitutes, beautiful crumbling hotels waiting to be merdified by the Hiltons of the world. Strangely, this is where I come to meet the last wheel of the Majorana cart: Pietro, Luciano's youngest son. He's led the life of a latter-day hippie: settling down meant opening a couple of bars in the Alps near the Austrian border (as far away from Sicily as is possible without leaving Italy, I note). He's now retired from that. All his life he wanted to be an architect. So now he's restoring a house, for himself, his wife, and his daughter to live in, one day.

Pietro and I drive out of Sanremo in a wreck of a car, past a sprawl of *abusivi* (illegal buildings) flanking greenhouses and agricultural patches. Artichokes and flowers are grown in fields overlooked by an ugly motorway, flying over the valleys on a concrete viaduct floating in noisy skies. But we're headed for the misty mountains, and soon civilization peters out—the province of Liguria is crushed against the sea by indomitable peaks. As we hit fog banks, the road becomes rougher, eventually thinning to a mere walking trail. This is where we park the car.

Nearby, there's a building site—the last house up the mountain. It's his pet project. A mismatch of seventeenth-, eighteenth-, and nineteenth-century structures,

the huge farmhouse has been left to go to ruin for over fifty years. He's been lov-
ingly fixing it up, using only original materials—the right mortar, the right bone
pigments—some of which have been used since the Roman times. The work's
been ongoing for three years; he tells me he doubts he'll ever finish.

The top floor is mostly done: There are no windows, but it's fully painted and has
a roof. Inside, there are no walls or stairs, but the balconies have been completed—
my God, the views! The little top room, carved out of the slanted roof, is to be his
bedroom, overlooking the entire universe. But nothing else—five floors arranged
over two houses—has a destined function yet. "We'll see what we'll make here,"
is Pietro's attitude. Plumbing and electricity will be added later, too.

Climbing over a small bridge linking the two houses, we enter a small chapel,
filled with rows of Formica chairs: He bought them from a cinema that went bank-
rupt ("Maybe we'll make a home theater in the chapel"). The altar is flanked by
two giant cacti that used to be in one of his bars ("I became attached to them").
There's no plan, no order, no priorities. None of this "we'll first finish this floor to
live in, then we'll do the rest"; none of "here will be the living room, there will be
the kitchen." He's going at it as a *whole*, expecting the house to impose its beauty
upon itself as it grows, determining its own functions: a new way of doing archi-
tecture. He clambers around ladders and scaffolding like a cat; in the dusk's failing
light, I stumble and almost kill myself a couple of times. I can see the beauty that
is struggling to emerge.

It definitely runs in the family, I think, recalling Ettore's 1932 masterpiece. This
house is the best insight I've ever had into Ettore's scientific mind, specifically his
1932 unfinished symphony. There are supreme pieces of work that just don't make
any historical sense—Stravinsky's *Rite of Spring*, or Ettore's little gem, for example.
They exist outside time, for no rational reason whatsoever. This house makes no
sense, either. And like the *Rite of Spring*, or Ettore's paper, it's beautiful in a crude,
primeval way.

"Now I know *all* the Majoranas are mad," I comment as we prepare to leave.
Pietro laughs with pride.

"My father was a *puttaniere*," says Pietro, his appraisal of Luciano setting the tone for
our conversation as we sit down at a seaside bar. It seems that Luciano introduced

the whole young male family to the joys of sex, in the form of high-class prostitution—and not only that. I'm not sure Ettore and Salvatore qualified (perhaps they were too close to their mother for Luciano to have played it safe.) Pietro tells me that his father was also a playboy—he still sometimes meets old ladies who tell him they could have been his mother. Pietro then reveals a bevy of other scintillating details. For example, he explains that Dorina and Fabio Massimo slept in separate bedrooms. He provides extensive descriptions of the sexual mores at the vineyard of Passopisciaro. He tells me scandalous stories concerning the fauna that frequented Maria's house in Rome. Here's a man who doesn't mince his words. I like that. He also has strong views on any matters related to Ettore.

He doesn't impose these views on anyone—and so he's totally devoid of arrogance. But he does state his opinions clearly and without doubt; a perfect antithesis to Ettore Jr.'s nihilism towards his uncle. *Of course* Ettore didn't commit suicide in 1938: Sicilians just *don't* do that; they love life too much. *Of course* Ettore's tragedy was solitude: the generalized human inability to communicate, "out of fear, the irrational fear of exposing oneself." *Of course* the cause of Ettore's collapse between 1933 and 1937 was the death of his father. It's *obvious.*

Fabio Massimo died in 1934, following a difficult period in Milan after some work-related problems in Rome. His death had a very strong effect on Ettore. For Pietro, there's not a shadow of a doubt that this was the leading factor in his crisis. Pietro says that if Ettore could have communicated with anyone, it would have been his father alone. Dorina was too shallow and unyielding; his siblings too distant and self-centered. Ettore's only source of human contact, of communication, was his father, Pietro tells me. And that lifeline disappeared in 1934.

Pietro lost his own father when he was four; he tells me that he hardly remembers Luciano. But his much older cousin Wolfgang-Fabio (son to Werner Schultze and Rosina) remarks that Pietro is just like Luciano, with the same mannerisms, language, and sense of humor. Before his hippie incarnation, Pietro was a timid kid with severe social problems. He says that the absence of his father was the key to his troubled years. When people talk about Ettore, they're often echoing themselves, but Pietro, in his own tragedy, may well be reflecting Ettore's feelings about the death of Fabio Massimo. Pietro's eyes wet a few times as he admits to his very unhappy early days. He may now be terminally cool, in his pink jeans and antediluvian leather jacket, but he doesn't bullshit himself—or me. He lived a very sad youth in the shadow of Luciano's death.

Pietro and I drink freely, and the conversation lightens up again. If Luciano's sense of humor was anything like Pietro's, as suggested by Wolfgang-Fabio, I think I'd like to have met him. Pietro tells graphic stories about bad behavior during his brother Fabio's wedding; he recounts even more obscene tales about the shooting of a film at Passopisciaro, scripted and directed by the Majoranas; and he provides a full account of how Luciano met his wife, Signora Nunni Cirino. He also paints an endearing picture of Maria and her unconventional drawing room at Via Salaria in Rome. Swearing was de rigueur—the kids loved it. One infallible regular was a famous painter who procured his models at the bus stop just outside. Imagine the pickup line "Can I paint you naked?" But it worked—and he did paint his models "after fucking them, and before introducing them to Maria's soirees."

Considering Pietro's unusual openness, I'm not surprised that he is the only person I met who raised the issue of Ettore's possible homosexuality. There is no hard evidence for or against it, but as with Segrè's syphilis, if it were true, it would explain a lot. Pietro tells me a story that would illustrate much about Ettore, assuming he was indeed homosexual. By necessity, it would have to be of the closet variety, and this would have distorted his genius beyond recognition.

Pietro recounts that at one of his bars in the Alps there was a piano. One day a tramp walked in, ordered a drink, and asked if he could play. It was during a quiet hour, so Pietro thought, *Why not?* The tramp sat at the keyboard, and the next thing Pietro knew, he was left speechless by an outburst of Rachmaninoffian prodigy.

The tramp returned many times, always displaying the same remarkable metamorphosis from alcoholic loser to musical genius. Eventually Pietro became curious. Who was this guy? A misunderstood Mozart? Another cinematographic David Helfgott? After asking around the village, he finally discovered the story behind the genius-tramp. He was simply a man who hadn't come to terms with his own homosexuality.

SEVENTEEN

Meanwhile, at Via Panisperna

❄ ❄ ❄ ❄ ❄ ❄ ❄ ❄

While Ettore languished in the throes of creative failure, bereavement, a bad stomach, and an unused cock (whether for lack of a he or a she), life went on at Via Panisperna. Belatedly, Fermi had seen Ettore's wisdom, duly pushing on to become the number-one expert in the field of neutrons and nuclear forces. And yet the Majorana curse was upon him: "Physics is on the wrong road" was Ettore's pronouncement when Amaldi force-fed him the happenings taking place at Via Panisperna. These events would become Fermi's terminal trauma, worse than his early scoop by Pauli. They concerned something of paramount historical importance: the fission of uranium. And the dark secret stored in the basement of Via Panisperna.

The incidents surrounding the discovery of the fission of uranium—the process behind the atomic bomb—could be used as a definition of *fuck up*. To understand how such a blunder could have happened at Via Panisperna in 1934, one must backtrack a few months and note that Fermi was then riding on a huge high. In such circumstances, people relax and lower their guard. The fuck-up can then detonate in their faces, unexpectedly and unchallenged.

The time was the academic year 1933–1934—just after Ettore retreated into his shell—and it could be called Fermi's annus mirabilis, in analogy with Einstein's 1905 and Newton's 1666. He'd been severely stuck in his own prejudices up until then, and apart from Ettore, so had everyone else at Via Panisperna. As late as

1932, evidence in Fermi's notebooks indicates that he didn't believe in the neutron. Fermi had been marooned: Perusing his notes is an exercise in humility. How could someone who'd be so crucial to a scientific revolution be so far off at first?

In October 1933, the seventh Solvay Conference took place, ostensibly to resolve the mystery of the nucleus. Part of a series set up by Belgium industrialist Ernest Solvay, these exclusive invitation-only meetings gathered the top minds in physics.* Fermi was invited. Ettore wasn't. But the congress was dominated by Ettore's theory of strong forces, as reported by Heisenberg. What Ettore perceived as his 1932 failure—his unfinished symphony—passed unnoticed. This congress could have been a moment of triumph for Ettore, but by this time he was no longer in public circulation.

One wonders just what Fermi's feelings toward Ettore's success were, but in any case they had the right effect: They spurred him on to shake off his obstinacy regarding the neutron's existence and to shift from atomic to nuclear physics. This unblocked his creativity and led to his best scientific period.

Fermi first set to work on his godson, the neutrino. In December 1933 he produced a landmark paper on beta decay, developing the tools still in use today for computing the probability of a beta transmutation, with the emission of an electron and a neutrino. To gravity and electromagnetism, Ettore and Heisenberg had added the strong force, capable of binding the nucleus together; Fermi now postulated a fourth force—the weak force—responsible for beta decay and the creation of neutrinos. With his theory he predicted a plethora of probabilities for "weak" processes: decay rates of isotopes, the inverse beta-decay process (later used to detect the neutrino), and others. The central constant in "weak theory" is nowadays called Fermi's constant.

"A fantastic paper . . . a monument to Fermi's intuition," was the judgment of a major nuclear physicist decades later. And Fermi was very proud of his accomplishment, as he enthusiastically explained it to the Boys congregated in the Alps for a spell of skiing that winter. "I could hardly sit . . . bruised as I was from several falls on icy snow," related Segrè. Needless to say, Fermi's paper, "Tentative Theory of Beta Rays," was rejected by the magazine *Nature* as "too remote from physical reality." This irritated Fermi a great deal.

---

* The only Italian until then to have been invited was Corbino (in 1924), and he didn't go.

Next, Fermi steered the whole research at Via Panisperna to *experimental* nuclear physics, much to the consternation of Rasetti, who felt more at home with atomic physics. Overnight, Fermi converted from a "neutron denier" into one of its greatest authorities. Ernest Rutherford, the leading nuclear experimentalist, wrote to him, "I congratulate you on your successful escape from the sphere of theoretical physics! You seem to have struck a good line to start with. You may be interested to hear that Professor Dirac also is doing some experiments. This seems to be a good augury for the future of theoretical physics." Thus Fermi embarked on the route that would lead him to uranium—and his biggest blunder.

In January 1934, the couple Irène and Frédéric Joliot-Curie accidentally discovered artificial radioactivity. As I explained earlier, isotopes existing in trace amounts in nature are unstable and decay into other elements with the emission of radioactive radiations: alpha, beta, and gamma rays. But by bombarding stable (i.e., nonradioactive) elements with alpha particles (obtained from naturally radioactive isotopes) the Joliot-Curies produced new, man-made radioactive isotopes. This was a dazzling discovery—yet another installment in the alchemical dream of transmutation. Yet a bad omen hung over the discoverers at once. "We are entitled to think that scientists, building up or shattering elements at will, will be able to bring about transmutations of an explosive type," said Frédéric Joliot-Curie in his Nobel Prize address. He went on to express his fear of a cataclysm should these transmutations "spread to all the elements of our planet"—perhaps the first recorded statement of nuclear Armageddon.

After reading the Joliot-Curies' paper, Fermi threw himself into the production of artificial radioactivity. The Joliot-Curies had used alpha particles as their bullets (which, recall, are made up of two neutrons and two protons); Fermi opted for neutrons instead. He was ridiculed by the experts (Otto Frisch labeled his experiments as "really silly") because the neutron sources available at the time were much, much weaker. But Fermi realized that neutrons were vastly more efficient in producing transmutations. Alpha particles are positively charged, and so are all nuclei; therefore, there is an electric repulsion between the two, forming a barrier to be overcome before the alpha particle can penetrate the nucleus and cause a transmutation. No such barrier afflicts the neutron, because it is electrically neutral. Neutron sources were weaker, but their bullets much more effective. So Fermi gambled on neutrons.

The trouble was that the Via Panisperna group had no experience whatsoever in experimental nuclear physics. To make matters worse, Rasetti, Fermi's main hand with experiments, was undergoing a crisis of scientific belief (the first of many.) On a whim, Rasetti had recently abandoned his experiments, to go on an extended holiday to Morocco "in pursuit of an elusive phantom, the hope of finding something, somewhere that would give him satisfaction and a sense of fulfillment," according to Laura Fermi.* Undeterred, Fermi jumped alone into the dark.

It is at this point that I have to reveal the nature of the treasure locked in the basement of Via Panisperna. It was inside a safe, and the key was in the pocket of Professor Trabacchi. For medical purposes only, inside the safe lay one gram of radium, that important radioactive element first isolated in large amounts by Madame Curie. It was worth about 700,000 lire. By comparison, the annual equipment budget at Fermi's disposal was 40,000 lire, already ten times the normal budget of other physics departments. It was out of the question for Fermi to buy this radium.

But radium gives off a gas called radon. Professor Trabacchi—reinforcing his nickname of the Divine Providence—happily allowed Fermi to pump the gas from the safe into his lab. With some clever adaptation, Fermi converted this radon into a good source of neutrons.[†] Fermi began bombarding elements with neutrons and by March 20, 1934, had achieved artificial radioactivity in aluminum. An urgent telegram was sent to Morocco, summoning Rasetti back to Rome. The Via Panisperna Institute had joined the nuclear race.

The Boys began a systematic investigation of the production of artificial radioactivity using neutrons. By midsummer, they'd produced no fewer than three-dozen new "artificially made" radioactive isotopes. But in the middle of so much mindless sweat, a drama occurred, requiring serious brain power. Accidentally, they hit on

---

* Rasetti was considered "exotic" by the other Boys (particularly Segrè) because he "read books" and traveled places. His schedules were also more irregular that the others', with the exception of Ettore. From this early crisis, one can guess that his Boyish attitudes were always a facade. He had a lot in common with Ettore, as we shall see.

† Radon emits alpha particles, which, acting on beryllium, produce neutrons. Usually radon is a bad source for alpha particles because it also emits gamma rays, which might mask the effects under study or disturb the instruments; but here it didn't matter, because Fermi was observing a delayed effect. He'd bombard the beryllium with alpha particles from radon, then remove the radon and use the beryllium as a neutron emitter.

one of their most important discoveries, an essential tool in nuclear physics to this day.

One day, Segrè and Amaldi were performing an irradiation, obediently doing as they were told by Fermi, who was about to depart for a conference in London. They obtained the required result—a given transmutation that led to a gamma-ray emission—which Fermi duly communicated during his conference. But the next week, Amaldi repeated the experiment, just to make sure, and was embarrassed to find a totally different result. Upon being told about the discrepancy, Fermi assumed human error and was incandescent. As Segrè put it, "On his return to Rome he scolded us for our apparent carelessness. We were not only unhappy, but confused, because we could find no fault with the various experiments that gave contradictory results."* Amaldi and Segrè were deeply hurt at having their ears boxed by the Pope, but even worse, Fermi—apparently not trusting them—set newcomer Bruno Pontecorvo on the case.

When Pontecorvo examined the problem, the mystery only deepened: He confirmed that the same target, when bombarded with neutrons, sometimes led to one type of transmutation, other times to another. He also observed the same phenomenon in other elements, not only in the case found by Segrè and Amaldi. Furthermore this was *not* the kind of randomness they'd come to expect from quantum mechanics, a randomness ruled by well-established odds and leading to repeatable experiments. This was total chaos without any pattern or rule. On Monday one result would be obtained, on Friday another. The outcome depended on which corner of the lab, by whom, and on which table the experiment was performed. Even quantum mechanics wasn't this crazy. It looked as if there were ghosts at Via Panisperna.

For months they puzzled, seeking a control, a "placebo," some sort of reproducibility in their experiments. They went through the dreadful game of perhaps this, perhaps that, perhaps something else altogether was changing every time they did their experiments. At some point in the cruel charade, they wondered whether their makeshift neutron source failed to keep a constant "power." To test their hypothesis, they encased the source in lead, in the hope of slowing down the neutrons and obtaining a pattern of changes in their irradiations. Slower alpha particles

---

* All material and quotes in this chapter are taken from the books by Laura Fermi, Emilio Segrè, Bruno Pontecorvo, and Richard Rhodes referenced at the end.

are less effective in producing artificial radioactivity; perhaps a similar effect was present with neutrons. But lead had no effect whatsoever on their experiments. Whereas something else—and, infuriatingly, they didn't know what—clearly did.

It was then that Fermi had a brainwave. One morning, during the examination season, when they were all very busy with chores, he chose on a whim to filter the neutrons with a block of paraffin instead of lead. No reason whatsoever, just a "why not?" To everyone's surprise, the production of artificial elements markedly *increased*! But this was deeply counterintuitive to them: With alpha particles, any filtering reduced their speed and therefore their ability to induce radioactive transmutations. Not so with neutrons, evidently. Why? Fermi went home for lunch plus a siesta.

When he came back, he had the explanation: They were being utterly stupid. Of course faster was better for alpha particles, because they had to overcome the repulsion from the nucleus. But neutrons felt no such electric barrier; and the slower they were, the longer they spent inside the nucleus, and so the higher their chance of causing a transmutation. (This is the sort of lateral thinking that would have caused no problem to Ettore—I note—but at least Fermi got there eventually.)

Fermi then realized that what was slowing down the neutrons was the hydrogen contained in the paraffin, and that some desktops in the lab were made of marble, others of wood, their hydrogen content thus being quite different. "Fantastic! Incredible! Black magic!" Fermi reportedly howled. In one stroke, two discoveries had been made: that hydrogen is the "moderator" of choice to brake neutrons; and that the slower the neutrons were, the more effective they were in producing nuclear reactions. The ghost at Via Panisperna had been exorcised—the experiments now made sense. And thus the Boys discovered slow neutrons, an absolutely essential tool in nuclear physics today, used in everything from weapons to nuclear-power stations.

What followed describes well the Boys' hormonal state at the time. Leaping from the building, screaming and shouting, they went to test their new toy with hydrogen-rich water, using the largest volume of that fluid at hand: the fish pond in the courtyard (see Figure 17.1). Rasetti raised salamanders there, but now, without a second thought, they jumped in, submersing the target and neutron source to repeat the experiment underwater. The caretaker came out, protesting:

"What the hell are you doing?"

"Experiments," was the reply.

The pond of the Via Panisperna Institute.

"But why here in the fountain? You'll scare the fish, maybe they'll die."

But unaware that they were now probably condemned to die of cancer, the fish swam on unperturbed; it was the men who yelled and frolicked in the water who seemed to be on the verge of heart attacks. Their experiment had worked: Water slowed down the neutrons even further, increasing manifold their ability to cause transmutations.

Later, they went to Amaldi's, whose wife had a typewriter, to compose a paper reporting their new discovery. Segrè recalled, "Fermi dictated, while I wrote; Rasetti, Amaldi and Pontecorvo paced the room excitedly all making comments at the same time." According to Laura Fermi, "They shouted their suggestions so loudly, they argued so heatedly about what to say and how to say it, they paced the floor in such audible agitation . . . that the Amaldis' maid enquired timidly whether the guests had all been drunk."

Next morning, Corbino, more levelheaded, foreseeing the industrial applications of the discovery, suggested they patent it. "I can't forget," recalled Pontecorvo, "the sincere, childish and vocal laughter with which the suggestion of Corbino was received, who, seeing Fermi and the others laugh so much, observed brusquely 'You are young, you don't understand anything.'"

And they really didn't. The only one who might have foreseen the implications of what had just been discovered was Ettore. But he had retired from the world.

During the maelstrom that was the Via Panisperna lab in that period, it was at some point decided to systematically bombard every element in the periodic table with neutrons, starting with hydrogen and helium and proceeding onward in an orderly manner. Tasks were distributed: Fermi did the calculations, organized the experiments, and led their execution; Amaldi prepared the "electronics" and the measuring devices; Segré was invested with the errand of procuring all the elements in the periodic table: hydrogen, helium, lithium, beryllium, boron, carbon, nitrogen, oxygen, etc. In some cases, it was trivial to track down a given element; in others, it was a laborious job. Eventually, Segrè unearthed a theology student who collected elements as a hobby (it takes all sorts to make this world). As Laura Fermi described it, "When in going down his list Emilio reached cesium and rubidium (two soft, silvery metals seldom used in chemistry), Mr. Troccoli [the theology student] got them down from the highest and dustiest shelf, saying: 'You can have these free.'"

The repetitive work began. To avoid contamination, elements would be irradiated by the neutron source at one end of a corridor and then tested for artificial radioactivity at the other end. If short-lived isotopes were produced, this involved a swift run down the corridor, and Amaldi and Fermi prided themselves on being the fastest runners. A Spanish professor, seeking to meet "His Excellency Prof. Fermi" was thus surprised to be nearly knocked down by Fermi, his lab coat flying out behind him, as he approached the sound barrier. "The Spanish visitor could not conceal the depth of his thwarted expectations," according to Laura Fermi. One also wonders what Professor Antonino Lo Surdo thought of these athletics.

In general, neutrons cause light elements to become lighter, say, by emitting an alpha particle (their A decreasing by three and their Z by two; see Figure 17.1). This is possible because their nuclear charge is not too high (their Z—or number of protons—being low) so the electric barrier for the emission of a positively charged particle is not overpowering. But for heavier elements, with their larger Z and so larger nuclear charge, this is not possible, so that when they are hit by neutrons, still *heavier* elements are produced. The archetype involves the neutron being captured by the nucleus, leading to an isotope with the same number of protons but an extra neutron; this then undergoes beta decay; i.e., a neutron inside the nucleus converts into a proton with the emission of an electron and a neutrino. So bombarding an element with Z and A eventually produces another with $Z + 1$ and $A + 1$ (see Figure 17.2).* It all made sense and was consistent with the nuclear theory worked out by Ettore and Heisenberg. And so they proceeded, element by

---

* In some cases, a second beta decay followed, leading to an overall transmutation going up the periodic scale by two protons. When you combine the effect of the two consecutive beta decays element (A, Z) finishes as $(A + 1, Z + 2)$.

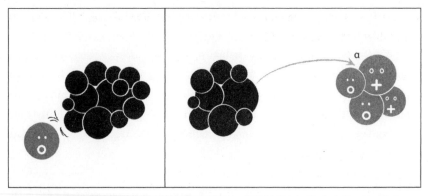

Figure 17.1: Neutrons running into light nuclei render them even lighter by stimulating the emission of an alpha particle. In the process, the nucleus gains a neutron but loses an alpha particle: Therefore its A decreases by three and its Z decreases by two.

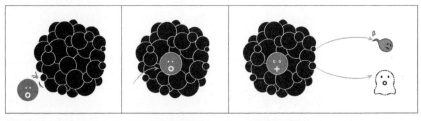

Figure 17.2: When bombarding heavy nuclei with neutrons, a different process is preferred. The neutron is first absorbed, leading to a surplus of neutrons over protons in the nucleus. The intermediate isotope thus created then undergoes beta decay: One of its neutrons converts into a proton with the emission of an electron and a neutrino. So bombarding with neutrons an element with numbers Z and A eventually produces another with both Z and A increased by one.

element, checking the predicted patterns, bored and slightly braindead. Until they finally arrived at uranium.

Element number 92—uranium—is the last natural element of the periodic table (i.e., the one with the largest number of protons.) Nothing with a higher Z exists in nature, and it was thus expected that by bombarding uranium with neutrons, element 93 and possibly 94 would be produced.* The political potential of such a discovery didn't escape Senator Corbino. If it all followed the usual molds, a new element or elements would be produced, not merely radioactive isotopes

---

* The issue hinging on whether one and/or two beta decays followed the capture of the neutron.

of existing elements. Someone would have to christen the new entries in the periodic table: Just think of the PR opportunity!

Shortly after the first irradiation of uranium with neutrons, Senator Corbino gave a speech to the Academy of the Lincei, with the king and the press in attendance. Displaying his brilliant oratory skills, he announced the discovery of the new element(s). Even though he acknowledged Fermi's "prudent circumspection," he stated that as far as he was concerned, the "production of this new element is already securely established." This set the press braying as usual. Add the reigning fascist regime and you shouldn't be overly surprised by the headlines: "Fascist Victories in the Field of Culture" mooed one paper; "In the Fascist Atmosphere Italy Has Resumed Her Ancient Role of Teacher and Vanguard in All Fields" cackled another. One paper went as far as to report that "Fermi present[ed] a small vial of the element 93 to the Queen." It was suggested that the new element be christened Mussolinium—but someone pointed out that, with the new element being so short-lived, this might be seen as an insult by the Duce. In the end the choice fell on Ausonium and Hesperium, derivative names for Italy, and thus suitably nationalistic.

The circus triggered by Corbino annoyed Fermi immensely: "I have seldom seen him in such a dark mood as after the speech made by Corbino" said Segrè later. First because Fermi didn't like publicity, then because he was very methodical and wanted to make sure nothing had been overlooked. He spent several sleepless nights while Oscar D'Agostino, a chemist recently lent to him by the Divine Providence, performed the relevant checks. Weeks passed by, with Fermi scared to death that a major mistake might have been relayed to the press. Then, as Fermi's notebooks reveal, he, too, felt sure they'd produced elements 93 and 94. As late as 1938, in his Nobel Prize speech, Fermi talks about this "discovery."

The "fuck-up" was complete. Later, Fermi would be bitterly aware that, in spite of all his achievements, he'd missed the greatest discovery of all. Before a picture commemorating him at Chicago University, Fermi once said, "It represents a scientist who was unable to discover fission." He also later admitted, "We didn't have enough imagination." For someone so traumatized by being scooped (and what it meant in the shadow of his dead brother), it must have been maddening. It's not as if he didn't have time to catch up with his error: It took almost five years for the issue to be resolved. And yet the writing was on the wall.

It's no accident that uranium is the last element of the periodic table to occur naturally. Everything with a higher Z is extraordinarily unstable. So when a neutron is captured by uranium, nature prefers to do something dramatically different: The uranium nucleus breaks into two much smaller nuclei (krypton with Z = 36, and barium with Z = 56) releasing a large amount of energy. Crucially, further neutrons are released during this "fission" process, capable of restarting the process *in an infamous loop called a chain reaction.* The rest is history: The atomic bomb, nuclear power stations . . . the nuclear age, in summary, is epitomized by fission. First done at Via Panisperna: except that no one noticed.

It would be easy to blame D'Agostino—responsible for the chemical tests—but that wouldn't be fair. D'Agostino's tests were standard issue and were carried out scrupulously. But the problem is that in examining the decay products of an irradiation, the chemist wasn't allowed to ask the general question, "What are they?" Instead he could only ask specifically, "Is it this; is it that?" And since no one knew the chemical properties of element 93, all he could do was to *rule out* that what had been produced were known elements. He checked the vicinity of uranium, for example elements 91 (protactinium), 90 (thorium), 89 (actinium), 88 (radium), and many others. He proved that these hadn't been produced; and thus the argument in favor of element 93 strengthened. He didn't check for elements 36 (krypton) or 56 (barium): Why the hell would anyone bother with that? It would contradict anything known in nuclear theory.

Well, in fact, someone did realize their fallacy straight away: Chemist Ida Noddack—by all accounts a Nazi and an unbearable person. In a paper published in a chemistry journal, she showed that the outcome of bombarding uranium could well be two pieces of comparable size, much larger than an alpha particle, but much smaller than "Mussolinium." She was ignored by Fermi, to whom she sent her paper. Segrè would later admit, "The possibility of the fission of Uranium escaped us even though it was signaled to us by Ida Noddack. . . . The reason for our skepticism is not clear to me even today."

Noddack was also ignored by all the other nuclear physicists. Fission simply wasn't a physical possibility according to the understanding of the nucleus encapsulated in Ettore and Heisenberg's theory of the nucleus. Ida Noddack discovered fission blindly where the others failed to see it because they had eyes. Ettore may have been the one exception. He was so disparaging of his own work—so aware of its limitations—that he may well have seen that the theoretical arguments were flawed . . . and the possibilities this opened with regard to uranium.

But Ettore was a recluse by this point, even though Amaldi kept him informed of these developments. It wasn't until 1939 that Hahn, Meitner, and Frisch set the matter of uranium fission straight. Just think of the historical implications, had Ettore brought fission into the world—with all its applications—as early as 1934. It's not surprising that conspiracy theories explain his disappearance by having him kidnapped by secret-service agents.

PAST   FUTURE

Figure 19.5: The Majorana neutrino doesn't know past from future, because it flows down the two conflicting arrows of time. Therefore the Majorana neutrino is its own antiparticle.

Ettore proposed that a neutrino is indistinguishable from an antineutrino, that particle and antiparticle are the same. He boldly asserted that a neutrino going backward in time is the same as a neutrino going forward in time. That what you call a neutrino is in effect a quantum superposition of particles going both forward and backward in time, in equal measure, or with equal probabilities, so that there's no asymmetry. And you thought Schrödinger's cat was schizophrenic, unsure about life and death. Ettore's neutrino is worse: a psychotic superposition of conflicting arrows of time made possible by the quantum ability to superpose opposites.

There is no arrow of time for a Majorana neutrino, because it contains within itself both time directions (see Figure 19.5). Majorana music might be defined as all songs that sound the same if played forward or backward. Odd music, one might think, but equally pleasing or displeasing in our world or in Amis's. Ettore's neutrino treats the two arrows of time democratically: It's ambidextrous in time. It simultaneously throws up and eats its meals. A Majorana neutrino flushing the toilet defies imagination.

The only change for a Majorana neutrino going backward in time is its handedness. A left-handed Majorana neutrino becomes right-handed. The mystery of the unoccupied states in Dirac's perception of the neutrino is at once solved. There are no empty slots in Ettore's theory. What you see is all there is: one slot for a right-handed neutrino, another for a left-handed neutrino. Make time go backward, and they just interchange.

The neutrino is still maximally chiral, but only because of the way it interacts with other particles. In beta decay, only right-handed neutrinos are emitted, whereas in inverse beta decay only left-handed neutrinos can be used as bullets. So there's handedness, but only in the way neutrinos talk to the world. Majorana's neutrino isn't a Dracula. If the left-handed neutrino gazed at a mirror, it would see

This designer clock illustrates well the interaction between time reversal and parity. It's twenty-four past three.

a right-handed reflection. Particle and antiparticle are the same, but the neutrino does have a mirror image.

Lepton number is not conserved in Majorana's theory, because the lepton number of the neutrino is zero. But this has never bothered anyone, because lepton-number conservation is a prejudice rather than a law. Majorana neutrinos can annihilate themselves. They also have several other oddities. But it's not easy to distinguish them experimentally from Dirac neutrinos. In fact, the constructions proposed by Majorana and Dirac are both realized in nature for other neutral particles which we know to be nonfundamental. The neutral kaon is a Dirac particle, whereas the neutral pion is a Majorana particle. The current score is 1–1, but this was a friendly match. With regard to the neutrino, the question is wide open.

Ettore admitted that an experimental choice between his option and Dirac's was not available, but this was twenty years before the neutrino was even detected. Later, "in the 50s and in the 60s the opinion was frequently expressed that neutrinos a la Majorana, although beautiful and interesting objects, are not realized in nature," as Pontecorvo wrote in 1982. But Pontecorvo went on to stress that things had changed, declaring the "question raised by Majorana" to be the "central question in neutrino physics." On March 26, 1938, Ettore disappeared, but not before leaving us this central problem. We're fortunate enough to live at a time when the controversy is about to be settled.

It was in this wild frame of mind that Ettore went to Naples in 1938: Having won his exalted position "by exceptional merit" with a paper proposing the Majorana neutrino, a particle for which the arrow of time doesn't exist—even though time flows, but in both directions at once. How befitting, given what was to come.

In Ettore's neutrino world, people both age and become younger at the same time. They first see the light of day soon after leaving a grave *and* a womb. Rivers stream from mountains to sea and from sea to mountains; but there's only one water, flowing without whirlpools or obstructions. The traffic goes both ways down the alley of time. Majorana neutrinos remember both their past and their future in a deluge of nostalgia and déjà-vu.

Suicide isn't a possibility in Amis's world: You have no choice but to eventually vanish into your mother. In Ettore's world, suicide is *and* is not an option. We're outside the duality of choice and imposition, ordering and disordering, positive and negative energy. When you look up at the sky, light leaves your eyes toward the stars but also springs from the cosmos. Tales begin with happy ever after, but also with a prince going hunting in the forest. Plot? You wanted a plot? Ettore's neutrino is like *losing the plot*.

Vanishing and appearing are one. Ships simultaneously leave wakes and cover their tracks as they progress along the seas. Welcome to Ettore's world, revealed just before he went to Naples—his last home on this side of reality. Assuming reality has sides.

TWENTY

The Quiet Before the Storm

In the first days of 1938, just after Epiphany, Ettore arrived in Naples, that exuberant city split between extremes of luxury and squalor, one quarter grotesquely peopled by very mature prostitutes, the next replete with ostentatious villas overlooking the gulf and surrounded by colorful flowers. "See Naples and die," goes the double-edged adage; in the old days Naples was the summit of the rich person's European tour—but also where many a genteel traveler died from disease or knifing.

Ettore went to Naples, and "he went to Naples willingly," as his sister Maria stressed in a TV interview many years later. He seemed upbeat, even enthusiastic, probably realizing that this was his last chance to rejoin ordinary life and break away from his family. He took up lodgings at one hotel, then moved to another, then yet another. He paid a visit to his boss, Professor Carrelli, and they instantly became friends, going together to buy his new office furniture (gracefully paid for by the university).

However, Ettore was not so keen on the formal pomp of an inaugural lecture. But the powers that be insisted: On Thursday, January 13, he had to deliver "La Chiamata" or the "Lectio Magistralis," as they called it in Naples: a lecture with no students in attendance but before the rector and the most important staff. As per custom, the newly appointed professor had to show off his ability to pontificate. The tradition was hated by all and was abandoned in the humanities only well into the 1990s. But in physics, Ettore was the last appointee who had to endure it. In

Ettore's registration form at the University of Naples. Health: somewhat delicate.

spite of his repeated requests for reserve, his whole family turned up. His lecture, on the foundation of quantum mechanics, was a success.

In the period that followed, we learn from his letters that he was slightly disappointed by what he found in Naples. "The Institute is reduced to the person of Carrelli, the old helper Maione and the young assistant Caccemo. There's also an old Professor of terrestrial physics, who is most difficult to locate." He tellingly misspells the name of the young assistant (Cennamo). I doubt he deliberately antagonized anyone, but the meager staff was comprised solely of experimental physicists, mostly of the classical persuasion, at that. He must have turned on his snobbery from day one.

He also discovered that he had only five students—not unusual in physics in those days; Rome had about twelve. And "it was in front of this audience of vaguely terrified youths that I presented Prof. Majorana," as Professor Carrelli related. "After which he started at once with his lesson." Looking at his lecture notes, we can't help feeling awe but also an element of the laughable. His students simply didn't stand a chance of understanding a word of it. He threw at his poor pupils an up-to-date course full of recent discoveries in quantum field theory and mathematical physics. It's an amazing course, the sort we provide nowadays—some seventy years later—as an optional subject for elite final-year students. I'm sure no one back then had a clue what he was talking about or could possibly have appreciated the value of his efforts.

Still, Ettore wrote that he was "happy with the students, some of whom seem resolved to take physics seriously." Given the size of his class, he lectured in a small

One of the last photos of
Ettore, circa 1938.

room—an *auletta*—situated on the ground floor of the institute. He entered from a lateral door giving onto a long, dark service corridor. With the students already installed in their seats, he arrived "silent and serious, not looking at anyone." He would climb up to the cattedra and without preamble start his lecture. His face "usually so gloomy then brightened up." The blackboard filled with formulae, his fluid, steady writing assisting an even sharper mind. Judging from his notes, the explanations he presented were original, modern, and crystal clear.

But often no one understood him—and he realized it. He then stopped, smiled, "forgot perhaps the great scientist that he was," and tried to help the students, seeking alternative, simpler explanations, even leaving the cattedra in his efforts to make himself understood. His students are unanimous in saying that although they couldn't follow him, "Professor Majorana was very generous and big-hearted . . . always tried to help those around him."

And so, after all those years locked in his room, Ettore settled into a routine away from his family. He led a quiet life of hard work and tranquility. In fact, there's very little abnormal in what is documented of this period. He wrote that he found Naples a bit odd because of the "scarcity of cars." (!!!!) His letters to the family are full of platitudes and details of practical chores. On February 23, 1938, he wrote to his mother in his trademark deadpan, "I have a discreet room; today they'll give me a better one overlooking Via Depretis from where I'll be able to see Hitler passing in three months. Are you cured from your cold? Perhaps I shall visit during Carnival." On March 9, so close to his final crisis, he wrote again to his mother, "Here the weather is very beautiful, ideal to sail in the gulf. . . . I shall perhaps visit on the weekend."

Chilling, if we reread this with the benefit of hindsight. He won't see Hitler passing on Via Depretis in three months, and he knows it. "Sailing in the gulf" will soon be dressed with a different, macabre meaning. At this point, he had already made his "decision by now inevitable." In fact, he had even confessed it to someone. He had also begun taking care of the logistics, all of a sudden withdrawing five months of salaries he hadn't previously bothered touching and asking his mother for his full share of their bank account.

On the surface, however, everything looked incongruously normal in Ettore's life, just like the quiet before a storm. Carrelli reported that he seemed very busy, "working on something absorbing, which he preferred not to talk about." He lectured in the morning on Tuesdays, Thursdays, and Saturdays but rarely put in an appearance at the institute otherwise. He worked hard, alone in his hotel room, or so everyone thought.

Then on Saturday, March 19, six days before D-day, he sent a bland letter to the family cancelling the visit he had promised the week before. He stated that he couldn't come because he had "business to attend on Monday at the names registry and elsewhere." He sent a telegram to ensure no one was waiting for him in the evening, and assured the family that he would visit "for sure" on the following Saturday. There is nothing unusual in his handwriting or style, and the family saw nothing strange in it. What a masterpiece of deceit!

His family expected him a week later. Which is, of course, when *it* happened, the run of events kicking off on the morning of Friday, March 25, when against his habits, Ettore visited the university on a day he didn't lecture. There he met one of his students, Gilda Senatore.

Gilda . . . a woman doing physics in 1938?

Oh, yes—I forgot to mention one little detail. In a deeply chauvinistic country and at a time when women weren't supposed to go to university, let alone study physics (recall Maria Majorana's experience), four of the five students who attended Ettore's course were women. And the last person he willingly met was one of them.*

These days, Dr. Gilda Senatore is best known for being the last person to see Ettore Majorana—or at least the last one he saw of his own accord—but the story of her life is interesting in itself.

When I meet Gilda in Naples, at the respectable age of ninety-four, she's still a force of nature: very talkative, extremely articulate, with remarkable lucidity. She's fond of purple prose and thumps her fist on the table with remarkable vigor (the first time she did it, I started with surprise). She holds my hand and looks me in

---

* The testimonies of Ettore's lecturing present in this section are Gilda Senatore's.

The profile of Dr. Gilda Senatore, at the age of ninety-four.

the eyes frequently. Apart from the inevitable eccentricities of a ninety-four-year-old, she is in superb form.

Born in Brazil of Italian parentage, Gilda spent her childhood shuttling between São Paulo and what she terms "the jungle," as the family followed her father, who worked in the mining industry and had to spend months on end in the wilderness. As we talk over lunch, I surmise that she still retains the fondest memories of these "undomesticated beginnings." She confesses to loving snakes even after reporting several chilling encounters*; she excitedly recounts how she'd vanish from the house to chase huge lizards in the forest with her Indian friends; she cherishes telling me, "I was breast-fed by an Indian woman, so I became a little savage." Per-

---

* Sample: "My mother was working on her sewing while seated on a stool with her legs slightly apart, when my father opened the door and shouted, 'Freeze!' He pointed his rifle at her and shot, the bullet hitting right in the middle of the eyes a huge snake standing upright under the stool in between my mother's legs."

haps due to these early experiences, she became a rebel as she grew up. This trait was to have a powerful effect on her destiny.

It must have been a shock to young Gilda when at the age of eight she was transplanted to Cava de' Tirreni, a tiny village south of Naples, and suddenly demands that she behave like an European were sternly made of her. She responded by adding an intellectual element to her brattish ways, making a nuisance of herself at school. She dropped ink on her immaculate white school uniform just to see the beautiful fractal develop, then asked why the fractal was the way it was, what light was made of, etc., etc., etc. . . . pestering the teacher with science questions he couldn't answer. Impertinences so unbefitting a young lady that her teacher punished her with a pair of donkey ears; to which she replied, *"O burro és tu,"* amusing herself to death when the poor man reported to her father her "peculiar fondness for butter."*

From the level of her questions, it's obvious that she was a physicist in the making. But with the air of the motherland, her father turned into a severe traditionalist, becoming a major hurdle for her talents. According to Gilda, he vented his view that "all women in my family can either be housewives or seamstresses." Gilda was not interested in either. And so determined was she to study physics at university that when her father threatened to disown her for her stubbornness, she retorted that she was ready to be thrown out. It can't have been an easy decision and all these years later, when I tell her the story of Maria Majorana's failed attempt to go to university, she becomes very agitated, stamping the floor for emphasis and thumping heartily on the table.

In the end, matters were made simpler for Gilda by the intervention of an illuminated uncle (a top doctor in the region), who offered her housing and paid her fees while she studied. She still had to give private lessons to fully support herself, but she did manage to attend university.

At university, Gilda informed Professor Carrelli that she felt honored to be offered the chance to study with such an eminent physicist, but unfortunately she couldn't care less about experimental physics, Carrelli's field of expertise. Theory, mathematics, and modern physics were her interests. Carrelli, half amused and half annoyed, passed her on to his colleague Professor Majorana.

---

* *O burro és tu* is Portuguese for "You're the one being the ass!" *Burro*—"ass" in Portuguese—happens to mean "butter" in Italian.

Which is where presenting another facet of Gilda's person becomes absolutely indispensable. How shall I put it?

I don't want to diminish the fact that young Gilda Senatore was very intelligent and had a strong, resolute character. That is a fact, and I couldn't stress it more. But it is politically correct crap when you go around talking to intellectuals who know and have studied her story, and they all fail to mention a very important little detail. A crucial piece of the story. *Absolutely essential.* It took an old-style Italian professor* to take me aside and break the news to me:

Gilda Senatore was a bombshell of a woman. The type that stops married men on the street even in the company of their wives, causes lethal car crashes with the casualties going to heaven still smiling—a female typhoon! All these years later, she still elicits comments such as the unnamed professor's: "As a man something inside you broke wild when you saw her; your heart started to race; you lost your head completely." She was, as they used to say in old-fashioned Italian, a *vampa*: a heat wave.

Is it un-PC to point this out? Being attractive hardly precludes being clever, for a man or a woman. In fact, Gilda herself is as proud of her juvenile beauty as she is of her intellect. When we met, one of the first things she told me was her "three measures" when she was young (bust, waist, and hips). But such is current Italy.†

Now we all know the troubles Ettorino had with women, the obsessions and complexes that vexed him. It must have made for some amusing viewing to watch the spectacle of Ettore entering the *auletta* only to find that four out of his five students were nubile young women and that one of them was terminally attractive. It's a measure of his remarkable talents that he managed to fill the blackboard with the intricacies of quantum theory while his mind may well have been elsewhere. And it's significant that the only student mentioned in Ettore's last letter to Carrelli is Sciuti, the only male, who by all accounts had been totally irrelevant to him.

---

* Who shall remain unnamed; but who isn't the obvious culprit—for those in the know.

† Recently a man was caught on the Milan-Lecco train staring twice at a woman, a transgression that was enough for the "victim" to complain that "he was undressing me with the eyes." The man got ten days in jail and a sixty-euro fine. In a culture that has so long lived with complete strangers staring at each other, one wonders if we shall soon also see Sicilian cannoli made with aspartame.

Given this state of affairs, it is not surprising that Ettore almost avoided Gilda Senatore. As Gilda comments, "He walked glued to the walls, like a shadow. Even the janitor commented on this." It's obvious that the whole department was affected by her, but no one more so than Ettore, with his patent need for love, repressed below so many layers of self-deprecation and low self-esteem. She must have represented the only plank of salvation possible in his sea of desperation. As his eponymous nephew said, "Love would have made all the difference."

Gilda was kind to him in a totally innocent way. But needless to say, he was far too shy to voice his feelings. He must have blushed, played the innocent love-struck boy, but no more. More generally he refused all contact and communication, with her or with anyone else. He was very solicitous in helping his students and her in particular, but he'd refuse anything in return, any form of affection in repayment for all he was giving them. "He was a very solitary man," Gilda sums it up. And so they hardly ever talked to each other. Until the day Ettore disappeared.

On the morning of Friday, March 25, 1938, Gilda Senatore went to the university to attend a lecture on terrestrial physics, after which, as she often did, she remained in the lecture room to study.

Suddenly she heard her name being called:

"Signorina! Signorina Senatore! . . ."

She raised her head and saw beyond the lateral door, still in the access corridor, Professor Majorana, that taciturn lecturer who never addressed any of his students and never turned up on the days he didn't lecture (of which Friday was one).

A bit taken by surprise, she didn't move, waiting for him to enter the room and talk to her. But he didn't budge either, expecting her to come to him. And when she didn't, he backed off slowly.

Something made Gilda get up at once. She ran to the doorway and then down the penumbra of the corridor. She quickly caught up with him and asked anxiously:

"What is it, Professor?"

Ettore stopped and looked at her fixedly. Something was obviously wrong. Gilda can't quite explain what, even after all these years and knowing now what happened next. But he just said:

"Here!" and placed a large box in her hands. "Here, Signorina! Will you please keep these papers for me?"

Startled, she took the box and began to seek an explanation. But he cut her short with a gesture, adding gently:

"We'll talk about it later."

And without letting her speak, he disappeared down the corridor, briefly waving her a goodbye.

She stayed there, frozen, wondering what to do with that box in her hands, while Ettore moved away, his reality already disintegrating. Later she will talk about this event very extensively: in TV interviews, in various writings, to me. And the major inflection patent in her tone is guilt: unexplainable, unjustified guilt, of the sort that is not appeased by others insisting that it wasn't her fault. Guilt for not having stopped him, for not having done something to help him. But, as she puts it herself, he wasn't susceptible to *receiving*. "Prof. Majorana liked to give generously, with simplicity and humility," she says. "But he would shy away as soon as anyone tried to give him anything."

I don't know what she thought she could have given him, when she reported these tragic moments so many years later. But even knowing full well how inaccessible he was, she still feels she could have done more, done something to stop him.

"Perhaps he wanted me to say something else to him, to extend a hand, to help him out."

Because he must have been wretched, miserable, depressed to the point of desperation. A drowning man flinging his arms out of the water one last time.

"But he didn't give me time."

And thus began the storm.

TWENTY-ONE

The Search Party

What could have happened to Ettore after he met Gilda Senatore? He posted his letter to "Dear Carrelli" and must have written the letter *a la famiglia* left in his room. He then had plenty of time on his hands until 5:00 pm, when according to the receptionist, he left the Albergo Bologna. It's a short walk to the port, from where the boat would depart at 10:30 pm. Did he have dinner? How did he fill these fraught hours?

The boat is about to leave and from the upper deck, some ten meters above the water, I appreciate what a drop it is. The water is filthy, awash with garbage and dead fish. Not a nice plunge, but one would likely wait for cleaner seas. The boat whistles noisily and slides away from the pier as I retrace Ettore's steps almost exactly seventy years on. Tirrenia, the company that took Ettore, is still in business. Admittedly, the trip is now made by a ferry carrying trucks and cars, but I still find it vaguely damning that it takes longer than it did in Ettore's day.

As we leave the harbor, we feast on great views of Vesuvius. How ironic that someone born in the shadow of Etna would leave reality under the quiet watch of another famous volcano. This one isn't as active as Etna, but is far more dangerous and unpredictable. Ask the people of Pompeii and Herculaneum who died buried by ash in AD 79, when the volcano most famously erupted. Since then, eruptions have been sparse but vicious; for example, in 1631, ash from an eruption fell as far as Istanbul, 1,200 kilometers (about 750 miles) away.

Vesuvius in the distance, as the Tirrenia ferry bound for Palermo departs from the port of Naples.

Fermi was quoted (by Carrelli) as saying, "With [Ettore's] intelligence, if he decided to disappear—or make his body disappear—he'd have succeeded." Mussolini's police chief, Arturo Bocchini, commented in response to Ettore's vanishing, "Corpses can be found, it's the living that disappear." But one of the most far-fetched explanations I've heard for the lack of a body is that, having sailed to Palermo and back, Ettore took the train to Vesuvius and threw himself into the crater. As we enter high seas, I note that he could have accomplished the same effect with much less effort. In front of me on the boat is a man-at-sea alarm. A sailor once told me that should someone fall overboard, whoever raises the alarm has to keep pointing at the spot where the person fell in. Otherwise, with no reference points, by the time the boat is turned about, it might as well have gone around the globe.

My mind wanders irrelevantly and mathematically as I consider what a truly two-dimensional setting the sea is. Getting lost on land, in no matter what labyrinth, is always one-dimensional and infinitely simpler. I see a mouse crossing the deck, and as I turn a corner, the full brunt of the sirocco pushes me off balance, making me slide on the wet deck. A puerile thought crosses my mind: Maybe Ettore just fell in by accident. In the distance, I see the lights of other ships. My mind wanders further.

And then I get a glimpse of what could have happened to Ettore on that anguished night, as he embraced his inner troubles. Maybe he didn't fall in by accident, but life might have intervened. It is known that the boat he took from Palermo was carrying an entire battalion returning from great conquests in West Africa. Imagine the din! In fact, I don't need to imagine, courtesy of a school trip, the teachers oblivious to their charges running riot and alternating screaming with serial vomiting (it only takes a critical mass of one to start a chain reaction, unlike with uranium).

There was I, trying to summon Ettore's tortured soul after he left Gilda Senatore, instead finding myself wondering if I should throw a couple of the noisy brats overboard. I might get some peace, and in the bargain test my theory that their bodies would never be found.

In an analogous environment, back in 1938, it would have been quite difficult to commit suicide (unless he jumped ship just to get away from the racket). Maria talks about a decision triggered by circumstance, "At dawn he was seen at the bridge," she said in Bruno Russo's documentary. "Dawn, as everyone knows, is the most delicate moment for those considering suicide. I am sure that Ettore threw himself into the sea, taking with him all the weight of his anguish and his atrocious doubts." But circumstances could equally well have conjured against it.

As the night unrolls, things quieten a bit on the boat, but never completely. I go to the bar and am gifted by serendipity: A TV set is blaring a moronic documentary titled *Did Hitler Have the Bomb?* Answer: no—after an hour of dim commentary and dramatic music, and just before the presenter is teleported out of the program. A funny coincidence . . . did Heisenberg and Ettore have more in common than we know? Heisenberg was certainly perilously close, at least conceptually, to the bomb, and this only a few years after that day in March 1938.

Back on deck, I realize what a gloomy figure I must cut: pensive and stark, staring at the sea. Maybe the insomniac brigade is worried I might be contemplating suicide. A girl comes out to smoke and waits to be chatted up. I move to the rear. What a sad bastard I must look, refusing to play the game of life, shouting and fucking, throwing up against the wind. I watch the wake for a long time, the cigarette butts flying past me into the night, like fireflies from Mars. In our world of the "normal," anyone who *thinks* is likely to appear suicidal. And yet, suicide or not, we will all be there one day, not just Ettore. We are all the same, only in different seasons.

Ettore was seen by Gilda on Friday morning, and then the factoids begin. Factoid number one: He boarded a ship to Palermo and then returned to Naples. This is known for a "fact" because, upon request from the family, Tirrenia located the tickets. Fact? Wrong. Factoid. Tirrenia merely *said* they recovered his tickets: No one actually saw them. Even today, in the age of computers, you should see the royal mess that pervades their ticket collection.

Then there are his letters. He posted one to Carrelli in Naples and left another in his room. The day after, in Palermo, he wrote a second letter to Carrelli on Albergo Sol letterhead, informing him of his change of heart. He rarely wrote on letterhead, so it looks as if he was trying to make a point. It's alleged that this letter proves he did arrive in Palermo, but again we are in the realm of factoids. It *is* his hand writing on a Palermo hotel letterhead, but one could imagine (admittedly forceful) ways in which he could have produced this letter without ever reaching Palermo.

If he was indeed in Palermo, then I can see why he sent a telegram to his Naples hotel stating, "Please keep my room." They hadn't seen him for a few days, and apart from the danger of giving his room away, they might have gone in, opened the letter to the family, and raised the alarm.

But on his alleged trip back from Palermo, something very close to *fact* did occur, at least if you believe Tirrenia's records. As chance would have it, Ettore shared a cabin with a Palermo University mathematician, Professor Strazzeri. An Englishman, Charles Price, occupied a third cabin berth. The latter was never traced, but the former answered an inquiry from Salvatore (enclosing a photo of Ettore). Professor Strazzeri replied that "it is my absolute belief that if the person who traveled with me was your brother, then he didn't kill himself, at least not before the arrival in Naples."

Upon closer inspection this, too, could be a factoid. Professor Strazzeri didn't talk to Ettore and could not, therefore, be sure that it was him. His account is also confusing; he says that the so-called Charles Price "spoke Italian like us, people from the South, and I assumed he was a shopkeeper or below, in short, someone without the refinement of manners that results from culture. . . ." How to explain this, since Price was English?

One possibility is that Tirrenia was mistaken, and Ettore wasn't in that cabin. A theoretically more elegant explanation comes from Sciascia, who argued a "third man" scenario. "Since Prof. Strazzeri exchanged few words with the man supposed to be Charles Price and none with the presumed Ettore Majorana," he wrote, "it's a reasonable hypothesis that the man who didn't speak, identified later to Strazzeri as Majorana, was in fact the Englishman; whereas the one he was told was Price was instead a Sicilian, a Southerner, the shopkeeper he appeared to be and who was traveling with Majorana's ticket." This explanation implies that Ettore remained in Palermo and gave up his ticket. It's contorted, but as Sciascia points out, it avoids the conspiracy of turning Price into a Sicilian passing himself off as an Englishman and following Majorana, "whence the myth of the Mafia trafficking in physicists as it traffics in women."

But there are worse factoids, such as the story of Salvatore and Luciano, already looking for Ettore, being accosted in Palermo by a stranger who told them: "I know who you are and who you're looking for; the friends of the friends are at your disposal." I don't need to translate the euphemism. A few days later, as legend has it, they received a complete report of Ettore's hour-by-hour moves in Palermo, sadly ending with his departure to Naples. "I am sorry to be unable to do more to help a man who has brought honor to our island," the stranger reportedly said. When I confronted Fabio with this story (presumably the report would still be with the family), his surprised face said it all.*

Alarm bells started ringing at top volume on Monday when Professor Carrelli sought out Ettore unsuccessfully and sent news to the family. On Tuesday—lecture day for Ettore—he had to tell the students that the course was cancelled. Imagine the gossip this must have sparked, the extravagant explanations the students would

---

* This Mafia tale is reported, for example, in Leandro Castellani's book—see References at the end.

have produced. Another of Ettore's pupils, Nada Minghetti, stated in an interview that when Ettore didn't turn up on Saturday, they were all delighted. It was a sunny day, and they were most happy to get a day off. It was only on Tuesday that the rumors started to spread, and gradually Nada realized that something serious had happened (I doubt Gilda Senatore ever thought otherwise). It was also on Tuesday that Salvatore arrived in Naples. He went to Ettore's lodgings at Albergo Bologna. It must have been eerie to enter his room: empty and silent; on the desk a letter "to the family." The room of a man who has abandoned the world.

Later, Luciano arrived in his car and the search began. Police and family worked independently but interacted. The family never believed Ettore committed suicide; the police, on the contrary, were certain he had. They placed a photograph of Ettore in the *La Domenica del Corriere* under the rubric *"Chi l'ha Visto,"* "Reporting Vanished People."* At first, Luciano and Salvatore thought he was still in Palermo, but then the evidence reversed to Naples. In addition to Professor Strazzeri's account, they received a report from the monks of San Pasquale a Portici near Naples. Then from a nurse who had treated Ettore and saw him between the palace and the gallery, coming from Santa Lucia. And then there was the priest of the church of Gesú Nuovo in Naples who swore he'd seen Ettore in his church in early April, begging to be admitted to a monastery.

All this got on Luciano and Salvatore's nerves—and thus the "search party" began. They scoured the Campanian countryside searching for Ettore in monasteries and peasant dwellings. These searches continued for over a year, from hamlet to hamlet and monastery to monastery. They didn't give up until Luciano, obviously distressed, one day ran over a little girl with his car. He was apparently very affected by this accident, and that's when he realized that enough was enough. The same level of commitment cannot be attributed to the police.

Initially, the police took a casual attitude towards the disappearance, believing it to be a clear-cut case of suicide. But then the relevant strings were pulled, rendering inappropriate their unconcealed lack of interest. Ettore's friend Giovanni Gentile Jr. implored his powerful father to intervene. Senator Gentile obliged and

---

* Ettore has been featured in *"Chi l'ha visto"* programs repeatedly; most recently on RAI, December 2008, as I'm finishing this manuscript. The original 1938 entry contained an appeal: "Ettore, return. Don't make us weep. Your family is waiting for you."

The architecturally rather odd Church of Gesú Nuovo in Naples, where Ettore allegedly made an afterlife appearance, requesting "an experiment of religious life."

went straight for the jugular: He contacted Arturo Bocchini. You should perhaps be made aware of what this meant.

All totalitarian states have a special police to "defend the state," spying on all activities, terrifying everyone, and torturing and killing anyone who fails to be cowed. In practice, this means nothing but legalizing and using the services of the usual psychopaths and other thugs to be found everywhere. Germany had the SS and the Gestapo; Portugal, the PIDE; Italy, the OVRO. Arturo Bocchini was head of the OVRO in 1938. He was the sort of man who could legally have anyone killed if he so wished; who could find out who had farted anywhere in Italy and beyond.

At the behest of Senator Gentile, Bocchini received Salvatore and ordered a dossier on Majorana to be opened. The contents, still on file today, are a masterpiece of deliberate incompetence. Regardless of Senator Gentile's efforts, clearly Ettore was a low priority for OVRO. For the powerful political police, it was merely a matter of humoring the Majorana family, in deference to Senator Gentile. The facts on file are vague and careless. They ignore claims that Ettore had

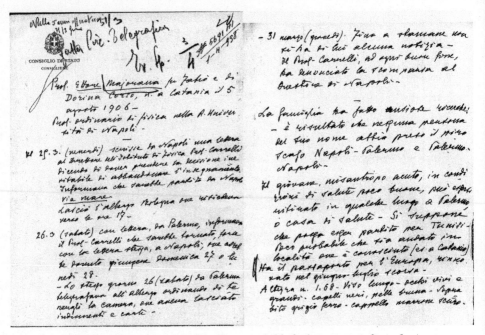

A couple of pages from Ettore's secret police file, remarkable for being so uninformed.

been kidnapped by foreign powers,* a matter falling squarely into their remit. Layers of blue, green, and lilac writing color-code, according to hierarchy, the reports documenting the diligences, or lack thereof, of the state police regarding Majorana. They don't even bother getting the dates right. It's blatant that no one cared.

Later, Fermi and Dorina petitioned Mussolini himself. Fermi's letter opens with *"Duce!"* the equivalent of "Führer."† Dorina's letter instead addresses the "supreme inspiration of all the Justice." Hypocritically, she claims that Ettore had "always been balanced and the drama of his soul and nerves is therefore a mystery." She insists he had no suicidal "clinical or moral precedents." On the contrary, given his

---

* An anonymous letter on file states that Ettore had been kidnapped by interests against the Italian state. It's hinted that he'd been completing an invention Marconi (recently deceased) had left almost in working order: a "death ray" that could be fired directly from Italy to Ethiopia. Apparently a cow had already been killed in a test strike.

† Fermi's letter is not in his handwriting. Whether it is a forgery or was written with his permission (or even dictated by him) remains a matter for debate.

life of study and hard work, she claims that Ettore should be considered a "victim of science." She notes that Ettore's passport would run out in August and entreats the Duce to step up the searches. "Excellence, [my son suffers from] a disease caused by noble studies, perhaps perfectly curable but destined to worsen without remedy if left untreated; only Your powerful intervention can decide on the fate of the searches and the life of a man."*

The folklore has Mussolini shouting to Bocchini, "I want him found!" extending a commanding arm and an accusative finger to emphasize the order.

Recall, if you may, that these events took place after the catastrophic vertex of fascism, beyond the point of no return, an alliance with Hitler already in place, racial laws being implemented, the aftershocks of a devastating war in Ethiopia still reverberating, only a year or so from all-out cataclysm. Bocchini must have thought, "Oh no . . . there goes that lunatic again." He wasn't going to endure another of His Excellency's eccentricities. As far as he was concerned, Ettore had obviously committed suicide.

A year later, on April 4, 1939, Ettore's file is closed, with no comment other than: "Archive and strike off." Two years later, Bocchini was poisoned by Mussolini for openly opposing the alliance with Nazism.

Gilda Senatore was ill for most of the rest of 1938, so she didn't regularly turn up at the university. She only found out much later that the university had lost all hope of seeing Ettore return and that it was preparing to replace him. All the while, she kept the box of notes Ettore had left her, waiting for him to return. But of course he never did.

She was so ill that she only graduated in 1939, a year after her colleagues. Her father, predictably, didn't come to her graduation. But her kind uncle did. Of her colleagues in Ettore's class, only Sciuti became a physicist, in Rome. Gilda also

---

* It's been claimed that Dorina's letter never made it to the Duce's office. Her letter isn't in the state archives, scrupulously kept during fascism. A Majorana entry can be found, but relates to a certain Maria Majorana—entirely unrelated to Ettore's family— who complained to the Duce about an estranged husband, only to be told off for being a slut by the official who investigated the matter.

wanted to pursue a career in physics, so she began work with Dr. Cennamo, the assistant professor whose name Ettore misspelled in his earlier letter about Naples. For a while, Gilda taught at the University of Naples. Then she became engaged to Cennamo and later married him.

I understand that she had a very tough life during the war, which she doesn't want to talk about. Painful memories she'd rather not mention, of the times bombs fell in Naples like rain. She's very much the survivor, but still a traumatized survivor. After the war, her career derailed. She never returned to her job at the university. Carrelli apparently didn't approve of married couples on his staff. She had five children; her youngest daughter is just like her.

At some point after she became engaged to Dr. Cennamo, Gilda mentioned Ettore's box of papers to him. It was the first time she had talked about it to anyone; you can see her dilemma, either way she was betraying someone. Alarmed, Cennamo ordered her to give him Ettore's box so that he could consign it to Carrelli. He was merely playing by the book: The *questura* had named Carrelli as the official custodian of Ettore's belongings in Naples. Reluctantly, her heart full of misgivings, Gilda gave away the treasure entrusted to her by Ettore. And whatever was in that box, trifling or exceptional, science or poetry, related to the neutrino or to love, it was never seen again.

In the background a fireplace is ablaze. It feels cozy, and we are in the Tuscan countryside. It's February 1990, and the old man who talks to the camera in Bruno Russo's documentary won't live much longer. He smokes a rollie and enunciates his words clearly, gripping us with his remarkable storytelling. His eyes gleam incongruously in his aged, wrinkled face. It's Giuseppe Occhialini, the famous experimental physicist, notable for the discovery of the positron (with Anderson and Blackett) and the pion. He met Ettore Majorana only once, quite by accident, when his ship stopped in Naples in January 1938, en route from Rio de Janeiro (where he was teaching) to Trieste. He had a whole afternoon to kill and paid a visit to Professor Carrelli. Then something happened which he would never forget.

As fate would have it, he nearly missed Carrelli, catching him just as he was leaving for lunch. But the professor received him cordially, postponing his meal and showing him around the institute, chatting about their current work. At some point a youth arrived, whom he took to be a student. "Dark eyes, dark hair," he remembers. He was flabbergasted to be introduced to Professor Majorana, whom he'd always admired as a genius. Eagerly, Occhialini had exclaimed, "But I've always wanted to meet you!"

In his Tuscan home the old man takes a puff from his cigarette and pauses to collect his thoughts. The ash at the end of his cigarette is getting perilously long and we can't help being distracted by it. Is it going to fall and burn him? But it doesn't, as the old man tells us *very carefully* how Ettore had responded to his enthusiasm. Smiling at the esteem bestowed on him by Occhialini, he said, "It's good you came now. Because had you waited another few weeks you would not have found me here."

The ash is now surreally long and crooked, perched at the end of his cigarette, and there's another pause in the old man's speech, this time much longer. His eyes dampen; he takes a series of puffs. At last he begins to talk again: He explains how he had *understood* at once, that Ettore's tone had struck a chord in his soul. He relates how when he was eighteen, he'd also contemplated suicide. And that he'd been moved to confess his own experience to Ettore.

When the camera that had closed on his face shows him again from afar we see that the ash from his cigarette is gone, doubtlessly fallen on him. But his jumper is not in flames as he informs us of Ettore's astonishing reaction to his confidence. Ettore said, "Look my dear Occhialini, there are those who talk about it and there are those who do it. For this I repeat that if you had arrived a few weeks later, you would not have found me."

Professor Carrelli was nearby doing something else, but he soon came back and they all had to go their separate ways. Occhialini tried to arrange to have lunch later with Ettore, but he had an appointment and couldn't accept the invitation. They said their irrelevant *ciaos* of the world and parted forever.

The old man spent that night on the deck of the ship sailing to Trieste, thinking over his moments with Ettore. Months later, he heard what had happened. And like Gilda Senatore and others who interacted with Ettore in his final period, his most obvious feeling is guilt, no matter how senseless and absurd. He remembers those moments "like one recalls when one met a girl and fell in love for the first time," a recollection made of "indelible ink," that stays with you forever and can be summoned at will. And he feels remorse.

By now he's using the ashtray, still chain-smoking. "We have all met complete strangers on a train who told us incredibly intimate things for no reason. We didn't meet for very long, but I felt that we were very close. And in that phrase 'there are those who do it' I felt the utter misery that was going through his soul." He feels like "a clown, someone who deigned to talk about suicide without truly meaning it to someone who did mean it." And when he heard the details of Ettore's *scomparsa*, he couldn't help thinking of Jack London's *Martin Eden* and his "death by water." The obsession with the sea mingled with that for death.

Like Martin Eden, Ettore was a good swimmer. His "watery embrace" would have been a slow and gradual diffusion into nature. If Ettore belonged to "those who do it," he went back slowly "by water" to the great dance of the cosmos, where time flows in both directions, as for his neutrino, back and forth, forth and back, to-ing and fro-ing. . . .

> Colors and radiances surrounded him and bathed him and pervaded him. What was that? It seemed a lighthouse; but it was inside his brain—a flashing, bright white light. It flashed swifter and swifter. There was a long rumble of sound, and it seemed to him that he was falling down a vast and interminable stairway. And somewhere at the bottom he fell into darkness. That much he knew. He had fallen into darkness. And at the instant he knew, he ceased to know.

PART TWO

AFTERLIFE:
THE DARK MATTER

TWENTY-TWO

Pagliacci

❁   ❁   ❁   ❁   ❁   ❁   ❁   ❁

We can only laugh at the ordeal Ettore inflicted on the university's administration, following the desertion of their "appointment by exceptional merit." Despite Carrelli's note to the Magnificent Rector, written within a few days of Ettore's stunt, it wasn't until August 16 that the university formally acknowledged his disappearance, confirming to the ministry that no salaries had been drawn since February. A replacement position in theoretical physics was requested and expeditiously approved by the minister to "support the teaching of the important discipline." A "removal order," however, was needed before they could begin getting through the red tape for a replacement. This was only signed, reluctantly, on December 6, quoting "abandoning office for more than ten days without justification" as the case for Ettore's removal.

Hilariously, the removal order was not accepted by the Court of Accounts, whose final approval was necessary to unblock the money allocated by the ministry. In Kafkaesque style, the court requested that Ettore himself sign an acceptance of his removal, confirming in writing that he had disappeared "without justification." The ministry and university countered in bureaucratic deadpan, stating that such a document, while of unquestionable value and importance, regrettably could not be provided. Wonderfully, they substantiated their declarations, not with an official police report, but with two newspaper clippings (one suggesting that Ettore disappeared to seek spiritual improvement). This was deemed acceptable by the Court of Accounts, but the decree still wasn't validated until September 1939 and even then committed finances only for a *temporary* replacement, lest Ettore return. That the war broke out in the middle of these shenanigans didn't help, but it's still extraordinary that Ettore's position in Naples was only readvertised circa 1950, more than ten years after his disappearance.

Martin Eden's body would never have been retrieved: "The lights of the *Mariposa* were growing dim in the distance, and there he was, swimming confidently, as though it were his intention to make for the nearest land a thousand miles or so away." A well-timed jump and Ettore would have joined the food chain long before anyone had a chance to find his body. What a simple explanation: Why, then, all this insistence on mystery, the belief that the body would necessarily have been restituted by the sea if he did commit suicide?

The leading "suicide theorist" is Bruno Russo, the man who produced by far the best documentary on Ettore (featuring the unique interview with Giuseppe Occhialini at the end of the last chapter). We have arranged to meet at Café Caprice near Piazza Duomo in Catania, and the city feels like an oven by 10:00 am when I leave the hotel. The day before, a 45° C heat wave triggered a state of emergency in Sicily, Calabria, and Puglia, with raging fires, power outages, and people collapsing everywhere. Blissfully unaware, I first spent too long on the beach and then climbed Etna, possibly with mild sunstroke. Now I feel awful, but it's too late to cancel.

I arrive at the café with Russo's book on Ettore in hand and do a few laps hoping to be recognized. A waiter approaches me, and I say I'm looking for a friend. He replies, smiling, "Maybe he disappeared," using the same verb everyone uses with respect to Ettore. I wonder if it's a jest—Ettore's picture is on the cover of the book. In fact, it's the picture edited from the photo of his father's wake that Fabio had shown me.

Finally someone comes toward me: "Professore Magueijo?" It's "Professore Bruno Russo." He cuts an odd figure, with a straw hat over a wet handkerchief falling on his forehead. He has a tumbled mop of wild gray hair and is wearing an incongruous combination of sandals and shorts with a formal shirt. At his suggestion we sit down inside, where the air conditioning should keep us cool; almost at once the lights go out. Waiters bring candles to the tables, making it look ridiculously romantic, but the temperature climbs ruthlessly, and soon we're both sweating profusely.

God, I hate people with a theory. They are all exactly the same. They *all* ridicule other people's theories on the grounds that they're based on a "biased selection of the data" or on "data that's clearly fake"; whereas, of course, their own theory is not. They *all* dismiss some facts and overemphasize others; then accuse everyone else of doing precisely that. They *all* recite a carbon copy of the same speech. In Russo's case, suicide acts as the editor of reality. Ochiallini considered himself a clown after his exchange with Ettore on suicide: He had high standards. The same modesty is sadly not present in Professor Bruno Russo.

His conversation reminds me of Michelangelo Antonioni films: reams of pretentious, French-imported, intellectual bullshit. He's one of those people who can't chat. When he takes the conversational floor, he keeps it for a quarter of an hour in a speech organized like a lecture. Things happen for three reasons: First [five minutes], second [five minutes], and third [final five minutes], and you're expected to reply in the same format. Interruptions are not tolerated. If something catches your attention, he brushes it aside until his lecture is finished, after which he says, "Tell me!" By which point you've totally forgotten, or couldn't care less, about whatever it was.

We talk about Ettore's personality and his passions outside science. He plays down Ettore's enthusiasm for Pirandello. For him, instead, Ettore's interest in Schopenhauer is the imperative clue to everything. A long lecture on pessimistic philosophy inevitably ensues, under extreme heat! Fucking metaphysics in 45 degrees Celcius, 113 Fahrenheit....

Schopenhauer's pessimistic philosophy, in the words of Russo, revolves around the scalene triangle of life, pain, and will. Life is equated with pain because it is also identified with will: blind, irrational will. But wanting something entails its absence and, thus, unrealized desire, which causes pain. Each desire come true only incites new unsatisfied desires, "unquenchable will" leading to further, inexhaustible pain. Thus man, by living and willing, is sentenced to eternal suffering. Worse: As an individual, he is condemned to a war without truce against other individuals, as their unsatisfied wills clash interminably, in a world full of strife and suffering. Is there any salvation on offer, then?

Not really. Unless you count negation of life as a solution. A Buddhist element now comes into the equation, as asceticism—the nirvana—is envisaged as, if not deliverance, at least the best man can hope for. If one is stricken by the abhorrence of being, and repudiates the world, the will, and all desires, then there's a glimmer of hope.

Is suicide then approved as a form of negating life? You will be thrilled to learn that Schopenhauer's answer is no, except when suicide is achieved by starvation. Suicide as a desperate emotional act is condemned (as an acknowledgement that the will is still throbbing.) But if deliberate death results from a negation of will, including the will to eat, then suicide is fully endorsed. Wonderful stuff, I think. Just what I needed to hear, in sweltering heat, drenched in sweat.

Yet the conversation gets still grimmer, "deep" in the intellectual Francomerde sense of the word. The café is walled in mirrors and someone takes a flash photo, making Russo jump with a start. I repress some laughter. He talks about the obscure spiral, illumination by death, suicide as the essence of existence. Unsurprisingly, his best-loved physicist is Ludwig Boltzmann, who hanged himself in the beautiful bay of Duino, near Trieste, while his wife and daughter were having a swim. I consider exhibiting my impeccable credentials: I was born in a region with one of the highest rates of male suicide in Europe. People think Sweden holds the record, but they only have the reputation (Ingmar Bergman is in part to blame). In fact, Alentejo's suicide rate in some categories is about twice that of Sweden's.

Nothing to be proud of, and come to think of it, I doubt Russo would have been overwhelmed. The Alentejo tradition is to hang yourself from a cork tree—not very classy. Besides, motive-wise, I imagine he'd be more interested in the Swedish variety of suicide: existentialist, first-world, well-fed (so that starvation can be an actual choice), driven by the abstract and the conceptual.

With great difficulty I finally steer the conversation into the concrete world only for derision to be poured all over me. He accuses everyone who looks at reality— the logistics of Ettore's disappearance—of trying to write a *giallo*, a thriller, which for such an exalted intellectual is like a venereal disease.*

---

* In Italy (as in France, Portugal, and Spain) a proper intellectual has to be impossible to understand by the masses. Anything "popular"—like a thriller—or which might reach a wide audience is scorned and put on the same level as football or bullfights. The real intellectual is several cuts above the populace, well entrenched in the hollow depths of verbal diarrhea.

"Laughable, laughable . . . that story of the passport he took with him, so that he couldn't have committed suicide. How do they know that he took it? Because they couldn't find it afterwards. Now I don't know how it was in 1938, but I can tell you that when I visited Majorana's house in Rome there was a mess, *but such a mess . . .* that I can't see how anyone can rule out that the passport has been there all along."

He also finds "laughable" the claim that Ettore was a happy person—*allegro.* "One day you crash a car, the next day you're *allegro.*" I tell him that it sounded like a hell of a party. He snorts. He won't admit any coloration other than dark on Ettore. It occurs to me that he's the one who is being simplistic, who is trivializing behind a veneer of existentialist intellectualism. But he prides himself on being the only author to have made a proper "map of [Ettore's] existential path," of his "free choices."

"There is coherence in people you can't avoid. Do you love the sea and then go on vacation to the mountains? Of course not: People are just not like that. Ettore had an obsession with the sea all his life, since he was a child. His life was meant to terminate in the sea."

Out of boredom I decide to play a little game: to pretend I have a theory. I'll edit the facts, accuse all others of doing what I've done, and claim I'm the only one who has accounted for a complete reality. The theory: that Ettore was a closet homosexual (something that only Pietro mentioned seriously). I disregard the evidence to the contrary. I then emphasize Ettore's religiosity (feelings of guilt, need for redemption, conservative morals, etc.) and his attachment to a domineering mother (I throw in some Freudian fakes for good measure). And presto: There you have another theory—the Ettore-is-a-poofter-in-the-closet theory.

Of course it's a joke, but he takes me seriously. He lists the full arguments against it: that Ettore was attracted to women, the letters to Gastone Piqué, and so on. I argue that the facts he mentions are obviously a cover-up. But where is your evidence? It would have been destroyed. But how? Blah, blah, blah. . . . This goes on for a while (I'm enjoying playing Ettorologist) when he finally shouts in exasperation:

"Then why don't you say that Ettore was a pedophile? Or that he was sexually attracted to ninety-year-old women?"

I nearly piss myself laughing. And he smiles too, for the first time. By this point I know I shouldn't like this guy, but I'm warming to him. I switch off my tape recorder; he relaxes and becomes quite a reasonable person. He forgets his existentialist agenda and we talk about the making of his excellent documentary, the remarkable interviews he so timely caught. By the time we part, the stiffness in our earlier conversation is gone. God, I hate people with a theory. They're such nice human beings otherwise.

The fact is that *all* Ettore theories explain only a subset of the data, but that has to be the case for any consistent theory, because the data is self-contradictory. So either you ridicule all theories—suicide, flight to Argentina, reclusion in a monastery, kidnap by aliens—or you embrace self-contradiction. I prefer the latter. There's nothing wrong with inconsistency when that's the only way to capture reality.

Russo's suicide theory is not ridiculous, but it *is* just a theory and therefore limited. It's undeniable that suicide was not alien to Ettore's psyche. Many years before those stark days in 1938, while walking with his friend Gastone Piqué along Via Cavour, Ettore reportedly sighed,"How sad it must be to live if one is very sick! I can't conceive that one continues to live if one is so ill." With his severe ulcer, these feelings could have resurfaced. But more serious are his tempestuous relations with the family, his mother in particular. Most suicides I have encountered are acts of raw revenge on those left behind. No one says so, out of charity.

TWENTY-THREE

A Pirandellian Intermezzo

✿ ✿ ✿ ✿ ✿ ✿ ✿ ✿

Suicide, however, is only one possibility. There's a good reason why the Suicide Party likes to discount the Pirandello influence on Ettore and overvalue his interest in Schopenhauer. Schrödinger was also a one-time Schopenhauer admirer: in the aftermath of the Great War, while he was undernourished and battling tuberculosis. But later the traitor changed his philosophical tack, once his health improved and he embarked upon the long chain of unbridled sex affairs that characterized his life. Without knowing it, Schrödinger's existence became somewhat Pirandellian.

Luigi Pirandello was born in the aptly named Agrigento suburb of Kaos in 1867. He is perhaps best known for his plays, but he was also a prolific writer of short stories and novels. His works feature a staggering array of delicious characters: monomaniacs, cuckold fetishists, blasphemers—a zoo often identified with the Sicilian human fauna. My favorite of his creations is the man who can't start a conversation without bursting out in laughter—for being unable *not* to conjure up the image of his interlocutor taking a shit.

In 1904 he wrote *The Late Mattia Pascal,* today compulsory school reading in Italy. *One, No One and One Hundred Thousand* is another of Pirandello's novels contemplating the same obsession: vanishing and identity change. It's not new, this thing of disappearing in Sicily. It's been said that many are, have been, and will be the mysterious disappearances in Sicily. Andrea Camilleri's *La Scomparsa di*

Patò recounts the story of an Easter performance of the Passion of Christ,* during which the amateur actor playing Judas disappears through the stage trap before hundreds of spectators, never to be seen again dead or alive. In Australia they call it "going walkabout"; in the hands of Pirandello, Sicily meets the outback.

Pirandello is the pinnacle of Sicilian literature, and there's nothing more Sicilian than cuckolding and vanishing. I doubt Ettore had much to do with the former, but he loved Pirandello. Did Ettore, in his final act, plagiarize his beloved author?

The character Mattia Pascal is a poor devil, the loser who gets the wrong end of the stick in a quintessentially Pirandellian multicornuto web. A crook has financially ruined his family; to get revenge, Mattia doubly cuckolds the crook (wife and niece). When the scandal breaks out, he finds himself forced by honor to marry the crook's niece, beforehand betrothed to his best friend (an idiot who's nonetheless dear to him). His new wife and mother-in-law then make his domestic life hellish. He gets by, playing the fool, because he's madly in love with his baby daughter, born in the meantime. But when by a stroke of fate his beloved mother and daughter simultaneously die, he can't take it anymore. Maddened by bereavement and directionless, Mattia leaves the town on the sly, intent on moving to America, Argentina perhaps, to start a new life.

But as he drifts he ends up in Monte Carlo's famous casino. Mattia has never played roulette or any other such game, and out of curiosity buys a book to learn the rules. Providence, having taken his loved ones, cynically begs to assist him as he tries his hand at the game. In an unlikely series of lucky strikes, he becomes immensely rich. A bit bewildered, Mattia decides to return home, determined to recover his lost property and put his life back on track.

On the train home, however, something ghastly happens: He picks up an abandoned newspaper and reads notice of his own suicide. The decomposed corpse of a drowned man has been found in his town, and since Mattia had gone missing and was known to be depressed, it didn't take much for the whole town to attribute the corpse to him.

---

* The *mortorio*, in Sicilian parlance.

At first he's indignant, and thinks of going back to confront those who so easily recognized him in a corpse. Then, with his pockets stuffed with cash, he has second thoughts. He hates his wife and still owes money to the crook. Here is indeed a chance to start anew. And thus he makes a clean break with his past by inventing an alter ego—Adriano Meis—with the looks of a German philosopher and an alternative biography passing through Argentina, conveniently having no living relatives in Italy. For months Mattia leads a nomadic life of pleasure, journeying throughout Europe, impersonating the wealthy Adriano as he moves from hotel to hotel. His identity is never questioned.

His own self, Mattia, is left to remain officially dead. Far away in his home town, no one contests this alternative reality.

But as time passes, the orgy of freedom begins to lose its appeal. Adriano feels lonely and in need of settling down but realizes that he can never own property or have a family. Meaning to acquire a sedentary life, he takes up lodgings with a Roman family, full of eccentrics who are happy to welcome him as one of them. He is contented until he becomes aware of the dodgy dealings that go on in the house: séances and spiritualism, thievery interweaving with the afterlife and illicit sex. His nightmare is compounded when he falls in love with his host's daughter, but because he's officially dead, he can never marry her. It's as if he kisses her "with the lips of a corpse."

When her much-hated brother-in-law robs him of a large sum of money, he finds himself in an impossible position. He cannot report the theft to the police, because he doesn't officially exist. But worse than that, by not doing so he mortally offends his loved one, who frantically wants the family to get rid of the rogue. She confronts him, but he can't explain himself. Desperate, Adriano goes for extreme measures. In a double-crossed deceit of reality he fakes the suicide of his alter ego, leaving his coat and hat by a Roman bridge, a suitable suicide note folded in.

The only flaw is that he takes all his money with him. A bit like what happened with Ettore. And in these pages, Mattia gives us what could be great insights into Ettore's "afterlife," should he have plagiarized Pirandello: his curiosity for what those left behind were now thinking, for what the press said about him and what the police might have done. "Bustle, amazement, morbid curiosity of strangers, hasty investigations, suspicions, wild hypotheses, insinuations, vain searches; my clothes and books examined with that consternation inspired by objects belonging to someone who has tragically died."

But, unlike (possibly) Ettore, Mattia isn't only faking a suicide: He's expunging his alter ego by suicide in order to resuscitate his real self, officially dead by suicide! He returns where he can never return. After his family has recovered from the shock of seeing a ghost, it emerges that his wife is now married to his best friend—whom he'd cuckolded in the first place. A glorious mess. Furthermore, a legal technicality would nullify the second marriage and force him to take back his ex-wife.

For a while, he toys with availing himself of the law, exacting a wild revenge upon those who were so keen to see him in that corpse. But then he meets the baby daughter born to his wife and friend while he'd been away. He remembers his own dead daughter, whom he loved so much, and his heart softens. In the end he lets the happy family be, for which, of course, he has to remain officially dead. Which he willingly does.

The only dark cloud for Mattia is the tombstone in the local cemetery with his name carved upon it. He regrets its existence not because he finds it gruesome, but because someone else really is dead underneath it. Perhaps the dead man has a family somewhere, not knowing where he is or what happened to him. God knows what extremes of despair drove the man in his tomb to commit suicide.

Mattia feels sorry for him, and on Sundays leaves flowers on his own tomb.

TWENTY-FOUR

Don't Cry for Him, Argentina

❀ ❀ ❀ ❀ ❀ ❀ ❀ ❀

The Argentine theory of Ettore's disappearance was put forward by Erasmo Recami and has its source in Carlos Rivera, a Chilean physicist at the Catholic University in Santiago. Picture the man, sitting alone at a table of the Hotel Continental in Buenos Aires, bored to death. To kill time, he scribbles a few mathematical formulae on the paper tablecloth.

"I know someone else in the habit of writing equations on the tablecloth," the waiter tells him, bringing him down to Earth with a start. "It's a client who comes here to eat, sometimes just for a coffee. His name is Ettore Majorana, and he was an important physicist in Italy a long time ago. But he's been in Argentina for many years."

It's 1960. Ettore would have been fifty-four.

With perfect idiocy, Rivera doesn't follow up the lead there and then. True, the waiter had no idea how to find this Ettore-who-wrote-formulae-on-the-tablecloth, but he could at least have noted the contact details of the waiter. As it was, when the clue was properly followed up later on, not even the waiter could be traced.*

---

* The material in this section (including all quotes) originates from Erasmo Recami's book and Leon Ponz's documentary (see References at the end.)

This was not the first time that Ettore had haunted Rivera in Buenos Aires. Some ten years earlier, in 1950, en route to Germany, he was staying at a small family hotel owned by a certain Signora Talbert. Again he was writing equations, this time on a pad of paper in his room. His notes concerned the Majorana statistical laws, and that's what he'd penned in glaring capital letters at the top of the page. Signora Talbert was hovering in the background doing the cleaning, when she suddenly exclaimed:

"Majorana? But that's the name of the famous Italian physicist who's very good friends with my son. They see each other all the time! He doesn't do physics anymore; like my son, he's an engineer now. He left Italy because he couldn't stand Fermi, an impossible fellow. He said he didn't even want to hear his name!"

A phone call from the son providentially discontinued the Signora's chattering and when she returned to the room she was uncomfortable, withdrawn, and refused to continue the conversation.

When Rivera seeks out Signora Talbert in 1954, he finds the hotel boarded up. The neighbors tell him that one night mother and son disappeared without notice and were never seen again. They were renowned anti-Perónists, and Rivera (along with many future sources) blames their disappearance on the "Perón purges." Rivera even speculates that Ettore could have met the same fate.

After Rivera's testimonies and their popularization in the press in the mid-1970s, sightings of Ettore in Buenos Aires spread like the plague. The leading proponent of the Argentinian theory became physicist Erasmo Recami, who found further traces of Ettore in Buenos Aires. Take, for example, the widow of the Guatemalan writer Miguel Asturias. Meeting Recami in Taormina, Sicily, Signora Blanca de Mora says without ado:

"But what's the big deal about Ettore Majorana? In Buenos Aires everyone knew him. I met him often at the house of the sisters Manzoni, descendents of the great novelist."

She alleges that one of these sisters, Eleanora, a mathematician, even dated him! But Eleanora is dead, and attempting to verify this testimony quickly becomes a comedy of errors for both Recami and Maria Majorana. Eleanora's sister, Lilo, has moved to Caracas, her address is found but then lost together with all traces of her. Then Lila, Blanca's sister, who was closer to the events, is brought into the correspondence. She admits to knowing Ettore, then takes it all back and refuses to put anything into writing, thereby making the story even more suspicious.

Recami broaches the matter with the authorities in 1979, but by now we are in the era of Videla's military dictatorship, and he encounters a wall of silence. The head of the Physics Department at Buenos Aires University promises to help "within the limits prescribed by the law." He then proceeds to deny the existence of the Hotel Continental in Buenos Aires, a noted landmark: a bit like denying the existence of elephants at the zoo while standing before their cage. Maybe by denying reality in such a blatant way, he was sending Recami a ciphered message: "Don't believe anything I say; my words are being censored."

Who was this Ettore Majorana? A figment of the imagination of frenzied women? A psychopath?

There is no Majorana in the Buenos Aires phone book, but in 2007 there were twenty-nine Maioranas, sixty-one Maioranos, two Mayoranas, ten Mayoranos, and one Maoirana. Some people can't spell. When I told Fabio that there were some one hundred Majoranas living in Buenos Aires he grinned mischievously before asking:

"And you think they might all be Ettore's children?"

As it happens, there is a district of Buenos Aires called Palermo. I stayed there when I visited. The city's Italian heritage is reflected everywhere, but they write *"ñoquis"* for "gnocchi," and *"al pomodoro"* becomes *"con tomatitos."* I've always been fond of these cultural distortions: In England anything cooked with tomato is "à la Portuguese"; in Italy *"alla Portoghese"* signifies leaving a restaurant without paying the bill. It looked as if there were far more women than men on the streets, but it could have been just my impression. There weren't that many casualties in the Malvinas/Falklands war.

The Argentinean Palermo is a *very* fashionable place to be, full of posh restaurants and designer-clothing shops, more so than the Sicilian Palermo. With Ettore in mind, I set off through the city, toward the center, seeking out the locations where he'd been sighted. First, the Hotel Continental. How could anyone deny its existence?

Founded in 1927, the hotel is housed in an impressive building with nine floors, on the wedge formed by a wide diagonal avenue intersecting a grid system, a sort of Broadway of Buenos Aires. Fernlike bundles of leaves hang from the cacarandá trees outside. It's hard to see exactly what the hotel would have looked like in Ettore's time, but it must always have been grandiose. Today it's too business-oriented for the likes of Ettore and me. The meeting rooms have names like Portfolio, Concepto, Estratégia. Artsy-fartsy decor attempts to hide the nature of the clientele. If you're polite to the servants, they're completely taken aback with surprise.

Down the road, at number 671, is the Ministerio del Interior (Registo Nacional de las Personas), just the place to ensure you know your identity. By the entrance a tramp is eating food laid down directly on the ground. Still further down the avenue is the Plaza de Mayo, where the eponymous *madres* congregated at one time to complain about the *desaparecidos*—the disappeared ones. Maybe at the Ministerio they were so busy forging identities for runaway Nazis that they lost the identities of other people. Ettore's mother should have come here to complain, too. The Catedral Metropolitana provides an island of peace—until you have to shake off the beggars and pickpockets. Tourists let pigeons shit on their heads for the sake of that perfect photo to send home.

I went to the hotel dining room, where Ettore had been sighted, and ordered an eccentric meal of oysters (hot and cold), a glass of white wine, and nothing else. There was I, a single man sitting at the Continental's dining room, writing formulae on a paper pad, like "Ettore" and Riviera did on the tablecloth more than half a century ago. Back in the 1950s, the tablecloths may have been made of paper, but nowadays they're fine soft leather—you'd certainly be thrown out if you wrote on them. *Discretion* was the word of the day among the waiters. They didn't look at what I was writing; the service was impeccable. (A lone man in a restaurant writing notes—they probably thought I was a quality-control agent hired to evaluate waiter performance; a concomitant display was put on.)

My mind wandered, assisted by the wine and a pervading lassitude. And that's when I had a perfect vision of who this Buenos Aires "Ettore" might have been. A poem was already lodged in my mind:

> *I'm nothing.*
> *I'll always be nothing.*
> *I can't want to be something.*
> *But I have in me all the dreams of the world.*

Fernando Pessoa is one of the most famous Portuguese poets, but he published almost nothing while still alive. Among the papers he left behind when he died, a large number of poems were found, attributed to a variety of what he called "heteronyms" . . . not exactly pseudonyms because he *was* them; not mere alter egos, but the result of a case of "multiple personality disorder" that entered the history of literature in a spectacular way. Three of these heteronyms wrote outstanding poetry: Ricardo Reis, Alberto Caeiro, and Alvaro de Campos; but there were half a dozen others we don't know much about because they were revoltingly mediocre. Pessoa would arrive at his local café and inform his friends who he was on that particular day. According to his friends, he didn't need to. It was obvious.

Some of the heteronyms "died" before Pessoa's death in 1935, but others were allegedly still "alive" and could have attended his funeral. Ricardo Reis, for example, went away to Brazil while Pessoa was still alive and was never heard from again. It can be said that Pessoa brought the concept of ghostwriter to a new level.

Like "Ettore" at the Continental, Pessoa sat alone in restaurants but asked for the table to be laid for two or three and for wine to be poured for his invisible guests.* I once saw a guy in a restaurant sitting alone at a table set for a large party, keeping up an animated if silent conversation, his gestures and soundless laughter revealing when a particularly witty joke had been told, the wine flowing freely. He wasn't disturbing anyone else, but the owner still called the police and had him ejected. I never patronized that restaurant again.

And that is where my mind drifted, in the dining room of the Continental, when I saw before me this "Ettore" of Buenos Aires. Truth and myth, life and death, the boundaries of the self, or the lack thereof. . . . Could the Argentine Ettore have been impersonated by a mythomaniac?

In José Saramago's book, *The Year of the Death of Ricardo Reis*, that Pessoan heteronym returns from Brazil after Pessoa's own death. One night, in his large Lisbon house, Ricardo Reis wakes to the sound of footsteps. He gets up to find a man sitting in his living room. He sighs with relief:

"Ah, it's you . . . what a fright you gave me! I thought it was a ghost."

Fernando Pessoa is smiling back at him.

The Pessoan irony of the Argentinian theory. . . . The possibility that somebody may have impersonated Ettore in Argentina, possibly as innocently as Pessoa impersonated his ghosts. But an impersonator choosing for an alter ego a man who has disappeared? What genius! Pirandello and Pessoa would both have been impressed.

------

* He drank it all in the end.

Something else not quite right about the Argentinean theory is the notion that Ettore became a "desaparecido" during the Perón purges. Everyone I talked to (right and left wing) said that this is a conspicuous anachronism. In the 1950s, during Perón's first reign, people were jailed, maybe even roughed up. Abducted and killed, no. It may have happened in a couple of cases, but no more. Not to the tune of 30,000 as during the Videla's "dirty war" of 1976 to 1983.

At the Buenos Aires book fair a "comrade in arms" of my anarchist days comes to meet me. We haven't seen each other for over twenty-five years.

"Remember me?"

" . . . "

At last I shout his name. We hug as he smiles, embarrassed. It turns out that it's not his real surname, just a decoy necessary during the dictatorship days but which he no longer uses.

Later we meet in Palermo. A massive thunderstorm is raging. I hear insistent mewling, but can't locate its source. Finally, through the corner of my eye, I glimpse a spot moving high above the ground. It's a black cat, moving with great purpose along ledges and roofs. They're lucky, I'm told. Lightning strikes the building opposite, creating a great commotion.

My friend has changed dramatically since the days he slept on the streets and sold fake jewelry in Algarve. He grunts with appreciation as he browses designer clothes flashing from the elegant shop windows. He's dressed in the same refined style, unlike me. I poke fun at him. He defends himself:

"That's *not* capitalistic! It wouldn't cost that much to dress the whole working class like this."

The night unrolls, the Malbéc flows. He tells me how sexually demanding Argentinean women can be (details provided) but that it suits him well. He's on a playboy-style vacation from his girlfriend of more than thirty years. I remember her very well from their Portuguese days.

I knew that they'd been in Portugal as exiles, that she'd been a "desaparecida" before they ran away. She'd been abducted by the state police and for many months he couldn't even find a record of her existence in jail, not knowing if she was dead or alive. I recall her telling me why she'd chosen Portugal for her exile. It was a matter of superstition.

One night they collected her and her mates from the cell, made them dress in rags, and shoved them into a truck. For a long while they were driven deep into the countryside. Finally they arrived at an old warehouse in the middle of nowhere. It had all the symptoms of the "standard," and they prepared for the merciful bullet.

Inside the warehouse there were seats and a movie projector. They were made to sit, the lights went out, and they were shown a documentary on Our Lady of Fátima, the Madonna who appeared to three little shepherds in Portugal in 1917. And then they set them free.

As the evening unfolds, my friend's mood changes. Comments on passing beau-. ties have now given way to stories about atrocities. I try to curb the Malbéc (he's had a triple bypass operation recently), but it's too late. At a furious pace he tells me of the people taken at night and what happened to them, of the castrations, rapes, sophisticated tortures; of the pregnant inmates who were made to have their babies, so that they could be taken away and given up for adoption to sterile military couples; of when it was logistically impossible to kill so many people, so they started dropping planeloads of them, drugged and loaded with sandbags, straight into the sea.

"Nowadays, if you want to insult someone in Argentina really badly, you call him *milico*—a military. It's worse than *hijo de puta*."

As his rage surges, his slurred speech spirals toward the fixed point, the thing he'd been thinking about all along. By now he has tears in his eyes. He wipes them with the napkin.

"You know they gave her electric shocks in the vagina. . . ."

I don't feel I have the right to ask him to stop, to spare me the unspeakable. I let his fury and tears take their course, torrents of gruesome details pouring out, feeling less and less capable of any distance. I find myself saying, in my Borat-style Spanish:

"That a curse be upon those fuckers!"

He pauses at long last. An incongruous smile appears on his face and he says:

"Well it is. After the pregnant inmates' scandal came out, all their teenage offspring started to ask for DNA tests, not believing it's their real parents."

We laugh, hysterically. I suppose that's the evolutionary purpose of laughter. Laughter, sometimes, is not funny.

The storm is now gone, and above the square we can see the stars in a limpid sky. Perched on the roof opposite I spot the same black cat I'd seen earlier in the evening, still broadcasting to the entire universe. Although I'm sure Ettore was never in Buenos Aires, except in a Pessoan sense, a powerful vision comes to my mind. In that vile world my Argentinean friend evoked, live forcible *desaparecidos,* people who disappeared without any real mystery, other than that of how certain "human" beings can accumulate so much filth in their minds. Ettore's story, in this context, is like a manifesto of freedom. A statement of liberty. "I shall be a *desaparecido* of my own will. A decision I've taken because I wanted to. Because disappearing infused sense into my life, in all the inscrutability that sense can have. Because it spared me the atrocities a normal life might have forced upon me."

The cat mewls high on the roofs; a beautiful orange moon has come out. They're lucky, no doubt. Black cats are lucky.

TWENTY-FIVE

They Thought the Sun Was Sick

❀ ❀ ❀ ❀ ❀ ❀ ❀ ❀

After Ettore's *scomparsa*, the neutrino—Ettore's brainchild—went on to become a major player in particle physics. Throughout the 1950s it engendered respectable research, earning many a Nobel Prize. First neutrinos were actually "seen," thanks to ghostbusters Reines and Cowan with their Project Poltergeist; no longer was the neutrino a mere theoretical expedient, as when Pauli first conceived it. Then the neutrino's parity-violating properties were exposed by Lee, Yang, and Wu. The neutrino was chiral. Reality was stranger than fiction.

From then on, new facts about the neutrino regularly emerged. But calamities never come singly. Electrons and neutrinos belong to a family of particles dubbed leptons (particles that don't partake in strong nuclear interactions). This family was about to expand.

Already in the 1930s, physicists examining cosmic rays (including Ettore's five-minute friend Occhialini) had discovered a background of particles no one had anticipated. Among these some behaved just like the electron-positron duet in the reactions they caused: They had to be leptons. However, the new particles were much heavier than electrons or positrons: about two hundred times so. They were baptized "muons," and it wasn't until the 1940s (largely thanks to old Boy Bruno Pontecorvo) that scientists realized that the muon was partnered with a "muon-neutrino." In fact, the muon and its muonic neutrino were perfect replicas of the electron and the "electron-neutrino" as far as their reactions were concerned. In the same way that a neutron could decay into a proton, releasing an electron and a neutrino, heavy nuclear particles produced in cosmic-ray showers could decay by emitting a muon and a muonic-neutrino. Fermi's weak force drove all these leptonic processes.

Later, in the age of giant particle accelerators, a third generation of leptons was exposed. In 1974, the group at SLAC accelerator, Stanford University, discovered the tauon, yet another facsimile of the electron, this time 3,500 times heavier. Having seen the pattern before, no one doubted that a tauon-neutrino should then also exist. But the tauon-neutrino wasn't actually detected until 2000, at Fermilab. The complete lepton family had now arrived.

Electrons, muons, and tauons had their corresponding antiparticles: the positron, the antimuon, and the antitauon. As for their neutrinos, that depended crucially on their being Dirac or Majorana, and the issue remained open. Dirac neutrinos have distinct antiparticles; Ettore's don't. The fact that neutrinos were now known to come in three types didn't change the outlook.

The neutrino has always lent itself to drama: You should therefore not find it totally unexpected that in this golden age of neutrino discoveries something catastrophic happened. It didn't come from particle accelerators or nuclear physics labs, but from the sun. If you're against nuclear energy, you should be campaigning against the sun: It is a nuclear device. And in the same way that nuclear reactors on Earth had been used to unveil the neutrino, the sun became a major asset for neutrino physicists. But something disastrous was found in the 1960s regarding the neutrino output of the sun.

The sun is nothing more than a natural nuclear power station, and as such it releases an abundance of neutrinos; indeed, the sun emits many more neutrinos than it does particles of light. Its nuclear reactions take place deep in the solar core, and the energy they release takes millions of years to reach the surface to be shed in the form of light. But because neutrinos are so tenuous and ghostly, they stream out of the sun in a matter of seconds, thereby offering us a real-time X-ray picture of the inner workings of the sun, ultimately allowing us to predict how its surface will shine in a million years.

In the late 1960s, John Bachall set out to compute the neutrino flux emanating from the sun, assuming basic nuclear physics and astrophysics. He then asked an experimental friend, Ray Davis, to detect these neutrinos and verify that it all checked out. It was a hard experiment because neutrino interactions are so rare. So to be certain the reactions he saw were not caused by mundane "ambient" radioactivity, Ray had to shield his apparatus very cautiously. He performed his experiments a few thousand feet underground in a mine in South Dakota. Only neutrinos can travel down that far; cosmic rays and other sources of interference are absorbed by the kilometers of rock above. Ray filled a container carried into these depths with six hundred tons of cleaning fluid, knowing that when a neutrino hits chlorine, the chlorine converts into radioactive argon gas, which he could measure.

John computed that Ray should see ten such events per week, all things considered (flux of solar neutrinos, probability of a conversion, and amount of chlorine.) Ray, instead, saw an average of around three: week after week, consistently, in clear defiance of John's prediction.

When John and Ray hit the world with this surprising news, they were met with disbelief. If you can't understand why, take a moment and consider the cheek of these two guys. One was 100 percent sure he knew how the sun worked and therefore how many neutrinos it emitted. The other claimed he could tell ten atoms apart within six hundred tons. And when he found three atoms instead, he was sure he hadn't missed the other seven. Bachall and Davis had found a serious discrepancy between theory and experiment and wanted the world to address it. Not very surprisingly, they got a lot of hostile response instead.

At first, scientists blamed the experimentalist (always the easier scapegoat on such occasions): There had to be something wrong with Ray's apparatus. The BBC Horizon programs of the 1970s make for some amusing viewing nowadays, when they featured reputable physicists lightly dismissing the whole thing, asserting that sooner or later the missing neutrinos would be found. But Ray Davis was sure of his experiment and stuck to his guns. He knew that if he had made a mistake and not properly shielded his device from ambient radioactivity, he'd be seeing an *excess,* not a deficit of neutrino events.

So then they blamed the theorist. Of course, John Bachall didn't know what he was talking about—how could someone be so sure about the workings of the sun, such a complicated thing? Physicists thus played at ridiculing the theorist; the more redoubtable proposed alternative models of the sun, providing concrete new calculations of the neutrino flux. But the sun really isn't that complicated and the variant models all gave the same neutrino flux, which seemed extraordinarily robust to details of solar structure. The only alternative solar models that produce fewer neutrinos were extreme and contrived, and were quickly ruled out by helioseismology—the study of solarquakes—an impressive method for obtaining information about the sun's internal structure.

In between these two groups inescapably sat the alarmists—the wacky cults fond of eschatological prophecies. Perhaps the sun is shutting down. Maybe we were seeing a third of the expected neutrinos because only a third of its usual nuclear reactions were taking place at the solar core. The energy currently being released at its surface came from reactions that took place a million years ago. If only a third of these nuclear reactions were now taking place at the Sun's core, then its energy output would decrease to a third over the next million years. Think of the dramatic implications on Earth! The end of the world is nigh.

The answer, however, was much simpler. The clue? The number three. The sun was emitting three times fewer neutrinos than it should. Not two, not five, but three.

The solution to the solar neutrino problem, as it happens, is vital to the controversy opposing Dirac and Majorana neutrinos—and the issue at its center concerns the neutrino mass.

But what is mass? The term has multiple meanings, both colloquially and in physics. Informally, *mass* means "amount of stuff" and weight is used to measure it. They're conceptually different: *Mass* means "resistance to acceleration" or inertia, whereas weight is the ability to excite gravity. But to make a long story short, the two concepts—mass and weight—can indeed be used interchangeably (the long story being Einstein's theory of general relativity).

Something else that can be confused with mass is energy, as you might have guessed from the famous equation $E = mc^2$. According to this formula, energy—the ability to change the world—is equivalent to mass. The exchange rate is very large (the square of the speed of light), but again relativity conjures up the unification of seemingly different concepts. By adding energy to an object, you imperceptibly increase its mass: Something hot is "heavier" than something cold. And, likewise, mass itself constitutes a huge reservoir of energy, as nuclear physics and the world of particle annihilations spectacularly illustrate.

Weight, mass, and energy are thus equivalent within the framework of relativity. However, there's one important distinction: "mass" versus "rest mass," or "energy" versus "rest energy." Motion is a form of energy,* so $E = mc^2$ implies that objects in motion are heavier or more massive than those at rest. In fact, that's why you can't accelerate an object to the speed of light: Its mass would become larger and larger, resisting the acceleration more and more. The mass of an object becomes infinite as it approaches the speed of light, and you simply can't summon enough force to push it all the way up to the speed of light.

Rest mass is the mass (or energy) displayed by an object even when it's at rest. The rest mass represents the object's "internal" energy, which it preserves when it's still, but which can be released by chemical or nuclear reactions. Particles with a rest mass can never move at the speed of light (they'd acquire infinite "moving mass" in the process). Conversely, particles that do move at the speed of light (like the photon) must have zero rest mass. A particle moving at the speed of light doesn't have infinite mass because its rest mass is zero, or rather the particle can't be seen at rest. Its energy derives from motion alone. Particles that move at the speed of light do so for all observers and are pure motion.

The world of particles is thus neatly divided between particles with zero and nonzero rest mass. Dropping the qualifier "rest," particle physicists talk about *massless* and *massive* particles. The former move at the speed of light for all observers, the latter for none. The proton and the electron are massive, the photon is massless. How about the neutrino? According to the "standard model" of particle physics, developed in the 1960s, the neutrino is massless. But what if the standard model were wrong?

It's historically curious that neither Pauli nor Ettore ever assumed the neutrino to be massless. If the neutron was proposed by Ettore as a "neutral proton," required to bind the nucleus, the neutrino was envisaged as a "neutral electron" necessary to enforce energy conservation in beta decay. The neutron mass is only slightly different from the proton mass, but it was quickly realized that the neutrino mass

---

* Energy is ability to change, and a moving car colliding with a stationary one results in evident changes for both.

had to be much smaller than the electron mass (for reasons to be made clear very shortly). Still *no one* assumed its rest mass to be zero.

Due to a number of mathematical problems, however, the new generation of neutrino physicists in the 1950s and 1960s posited that the neutrino moved at the speed of light and that its rest mass was zero. The only voice of dissent was Bruno Pontecorvo. He may not have been given a nocturnal induction exam by Ettore, but he surely came from the broadminded environment of Via Panisperna. There, no one had taken the neutrino to be massless.

But in 1967, Pontecorvo proposed something brutally insane, crazy beyond belief. At the time it was already established that neutrinos have "flavors." This is a technical term in physics (not gourmet cuisine) referring to the three different generations of leptons discovered. The electron, muon, and tauon neutrinos can be seen as different "flavors" of the same particle. Pontecorvo knew that quantum mechanics is intrinsically odd—the Lord is malicious—and that sometimes different types of quantities we're accustomed to measuring simultaneously become incompatible in quantum mechanics, ruled by mutually exclusive uncertainty. Position and velocity comprise one example, as I've explained before. If you know the exact velocity of a quantum car, you can't know where it is (it must be in a quantum superposition of being at all locations). If you know its position, you have no idea of its speed. So position and velocity in quantum theory are "observables that don't align": They can't be measured at the same time. *Pontecorvo proposed that mass and flavor do not "align" for neutrinos.*

This meant that if you knew the flavor of a neutrino, you couldn't know its mass (you must be in a quantum superposition of different masses); if you knew its mass, you couldn't know its flavor. Such a "mixing" of flavors and masses had a radical repercussion. If an electron neutrino is a superposition of different mass states (with well determined ratios), these masses will move at different rates through space. Thus the amount of the different masses will change as the neutrino progresses in space and time. And as the proportions change, so does the neutrino flavor: an originally "sweet" neutrino quickly turns "sour," then "sweet" again—and so on, in a phenomenon called "flavor oscillation." An electron neutrino oscillates into a muon and tauon neutrino as it moves, due to its funny kind of mass (see Figure 25.1). The rate at which the oscillations take place depends on the difference between the various mass states and something Pontecorvo dubbed the mixing angle. For typical values, the oscillations complete a cycle every few kilometers.

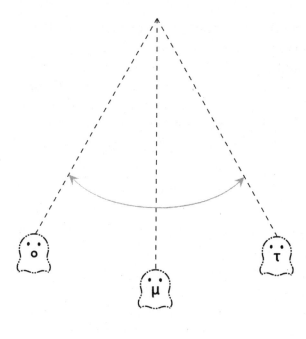

**Figure 25.1:
Pontecorvo
oscillations. As the
neutrino moves in
space, it oscillates
between being an
electron neutrino, a
muonic neutrino, a
tauon neutrino, then
back again.**

For a while, physicists continued to play the blame-the-experimentalist, blame-the-theorist, blame-the-sun game regarding the solar neutrino puzzle. But in the 1980s and 1990s, a Japanese group studying neutrinos produced by cosmic rays in the atmosphere found a similar discrepancy. They were also seeing fewer neutrinos than they should. The particle created by Pauli to account for missing energy was itself showing a knack for going missing. And when a larger Japanese experiment confirmed yet again a generalized neutrino deficit, physicists realized the game was over. It was time to start blaming the particle itself for the anomalies.

It was at this point that Pontecorvo's idea—largely disregarded up until then—began to gain respect. Pontecorvo had suggested that not only did the neutrino have a mass, but also that its mass states didn't align with its flavor states. As a result, an electron neutrino oscillated into a muon neutrino, which oscillated into a tauon neutrino, and finally back to an electron neutrino to restart the cycle. Ray Davis's solar neutrino experiment was only sensitive to electron neutrinos. No one had objected because the sun can produce only electron neutrinos in its nuclear reactions. But what if neutrinos oscillated? Two-thirds of the emitted electron neutrinos would on average be lost to muon and tauon flavors as they moved toward us. This would solve the solar neutrino problem. It was just a matter of putting

Figure 25.2: The (electron) neutrinos produced by the sun oscillate en route to Earth. Since the oscillation period is a few kilometers, and they came from a much broader region at the solar core, at any place on Earth we see on average a third of the total solar neutrino output in each flavor: electron, muon or tauon.

three and three together. Three generations or flavors; three times fewer neutrinos seen (see Figure 25.2).

It was a disarmingly simple solution to the solar neutrino conundrum, but there was only one way to verify that it was indeed the correct answer: to devise an experiment that could detect all three types of neutrinos. SNO—Solar neutrino observatory—started running in 1999, three kilometers (almost two miles) underground in a mine at Sudbury, Canada. It's now the cleanest environment in terms of ambient radioactivity in the whole solar system. You need to shower and decontaminate yourself thoroughly before they'll let you anywhere near. It was in such an exclusive location that physicists waited for ticks in their detectors, signaling the catch of another neutrino *of any flavor* coming from the direction of the sun.

The answer provided by SNO was unambiguous: Taking into account all types of neutrinos, the sun was in fine health, thank you very much. John Bachall's calculation was correct: The flux of solar neutrinos was as predicted. And Ray Davis was measuring things right, too. He just wasn't seeing all the flavors that issue from the sun because he had restricted himself to electron neutrinos. The sun creates only electron neutrinos in its reactions, but it just so happens that Pontecorvo's wild guess hit the spot. Neutrinos have mass but their mass states don't align with their flavors, and so the flavors oscillate.

As a result, a third of the solar neutrinos arrive on Earth as muon-neutrinos, and another third as tauon-neutrinos. There are more things in heaven and earth, Horatio, than are dreamed of in your philosophy. And this one couldn't be more relevant for the outcome of Ettore's story.

TWENTY-SIX

The Sign of the Beast

✤ ✤ ✤ ✤ ✤ ✤ ✤ ✤

It may seem inconsequential to give the neutrino a rest mass for the sake of saving the sun, but in reality the neutrino mass is crucial to Ettore's theory. Should the neutrino be massless, then the Majorana neutrino would be observationally indistinguishable from its alternative, the Dirac neutrino. Were it not for the neutrino's inability to move at the speed of light, Dirac versus Majorana would be a matter of taste, like supporting different football teams or belonging to different religions.

Recall that the controversy hinges on how the neutrino reacts to Amis's backward world—its behavior under a reversal of time's arrow. Dirac's neutrino in Amis's world becomes a distinct particle, the antineutrino. Ettore's remains the same because Majorana neutrinos already experience a quantum superposition of the two arrows of time. If the neutrino were massless, it would move at the speed of light, and time doesn't pass for such particles. The faster something moves, the slower time flows. It grinds to a halt for particles moving at the speed of light, so that massless particles like the photon live outside time. If the neutrino were massless, it wouldn't feel the flow of time, and there could be no distinction between indifference or not to time reversal. The neutrino *must* be massive for there to be antagonism between Dirac's and Majorana's proposals.

That the neutrino mass doesn't vanish was proved incontrovertibly by the various experiments that established neutrino oscillations (a process that also happens in time, and so requires nonzero rest mass.) In 2002, Ray Davis and Masatoshi Koshiba (representing the Japanese group) won the Nobel Prize for this discovery. The path toward the detection of Ettore's neutrino was now open.

As early as 1935, Maria Goeppert-Mayer realized that beta decay could exist in a form where two coincident decays happened inside a nucleus. Two neutrons inside the same nucleus could simultaneously, and as part of the same process, convert into two protons releasing two beta rays and two neutrinos. The process is very unlikely and Goeppert-Mayer computed its probability using Fermi's theory of weak interactions (the one famously rejected by *Nature* magazine). She found that it was exceedingly rare because it involved two simultaneous interventions of the weak force. The rare decay was unimaginatively christened "double beta decay."

Because the process is so much rarer than single beta decay, it can only be observed if single beta decay is not possible for a given nucleus. If single beta decay can occur, it completely swamps the double version, releasing beta rays much more copiously. But for some nuclei, single beta decay is forbidden (say, by the energy available or by the need to conserve spin), while double beta decay is allowed. One then has a chance of observing the rare process without the loud noises made by the beta stampede of its more frequent version. Double beta decay was finally observed in the laboratory in 1986.

Barely had Ettore left the world when in 1939 scientists realized that another type of double beta decay was possible. Usually one expects two neutrinos to be emitted along with the two beta rays. Wendel Furry proved that it was possible for such a double beta decay to occur without the emission of any neutrinos *provided the neutrino was of the Majorana variety*. "Neutrinoless double beta decay" is effectively a double beta decay process in which the two neutrinos that would ordinarily have been emitted annihilate each other. It can only happen if the neutrino is Majorana *and* has a mass. In fact, the probability of neutrinoless double beta decay is related to the neutrino mass according to a well-established formula. The larger the neutrino mass, the more probable is neutrinoless double beta decay.

But this is killing two birds with one stone! If neutrinoless double beta decay is observed, we will know that the neutrino is Majorana *and* what its mass is. By contrast, Pontecorvo-style oscillation experiments give us only the mass *differences* between the mass states plus an angle, quantifying their misalignment with the

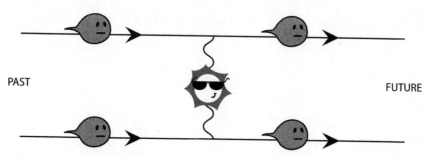

Figure 26.1: A Feynman diagram depicting the interaction of two electrons, via the interchange of a virtual photon.

flavor states. Double beta decay makes Ettore's idea a choice method for weighing neutrinos, explaining its current popularity with experimentalists.*

To understand interactions between particles, and the rare type of decay that may reveal Ettore's neutrino, we must once again appeal to Richard Feynman's lurid imagination. Not only did he propose that antiparticles are particles going backward in time, Amis-style, but he found an inventive and pictorial way to explain forces and decays. He invented the concept of *virtual* particles, promoting them to the intermediaries of interaction. He also crafted a useful tool for visualizing quantum processes nowadays called Feynman diagrams.

According to Feynman, when two electrons come together and electrically repel, they do so by exchanging a virtual photon; that is, a photon that's not quite real—*virtual light*. This is the kind of dark illumination a sun wearing sunglasses would shed: not "free light," not light you'd ever see. Rather, it's an obscure sort of light that is permitted only by the fuzziness of quantum mechanics, allowing energy to fluctuate at random for short periods. The virtual photon emitted by the electron isn't allowed by the conservation laws of classical physics and should never have been emitted. But quantum mechanics turns a blind eye . . . if it's not for too long. Long enough, however, for the virtual particle to carry energy and momentum between the two electrons, which thus repel (see Figure 26.1).

---

* Naturally if neutrinoless double beta decay is *not* observed, all we can say is that, given the sensitivity of the experiments, *if* the neutrino is Majorana, its mass must be smaller than a given figure.

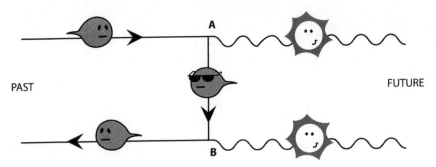

Figure 26.2: The Feynman diagram of an electron-positron annihilation (this is a refinement of Figure 19.3b). Not only does the electron pull a U-turn down the stream of time (to become the positron it annihilates), but it becomes a virtual electron while the photons are being emitted at A and B.

The Feynman diagram of an electron-positron annihilation is also enlightening. As I said earlier, in the Feynman picture of the antiworld, an electron-positron annihilation is like a single particle pulling a U-turn down the stream of time. If a positron is an electron going backward in time, an annihilation is like an electron reversing in time, providing you with the positron you thought you'd started with. Under closer inspection, however, the process involves this "single particle fits all" becoming a virtual particle for a short while (see Figure 26.2). The annihilation releases two photons, at A and B. Traveling between these two points we find a virtual particle, not permitted by classical physics. It can be seen either as a virtual electron travelling from A to B, or as a virtual positron travelling from B to A. The virtual electron is like a moment of "neutral" before the electron engages the full reverse down the stream of time, to become the positron it annihilated.

Virtual photons explain the electric forces between charged particles, and virtual electrons can appear in the Feynman diagrams of annihilations and other quantum processes. And many other virtual particles explain *all* types of forces, decays, and interactions, including the weak and strong nuclear force. If you think this is crazy, take note of the following: Feynman diagrams can be rigorously and mathematically derived from quantum theory; *and* they can be employed to predict particle interactions to a ridiculously high standard. Some particle physics quantities have been predicted to one part in 10,000,000,000 by Feynman's calculations.

In particular, Feynman diagrams can be drawn to explain beta decay. In beta decay diagrams, if the neutrino is seen à la Dirac, it must have an arrow like the electron, to distinguish between particle and antiparticle. For lepton number to

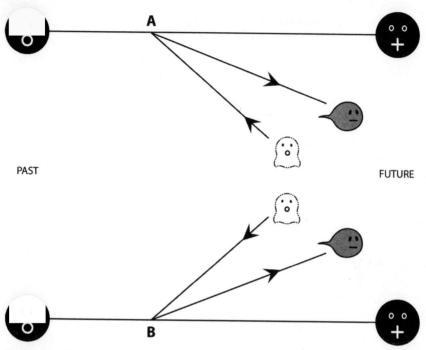

Figure 26.3: Double beta decay with neutrino emission is just two disconnected diagrams for beta decay happening at the same time. If the neutrino is Dirac (as in this figure) it has a definite arrow and so a clear particle-antiparticle nature. Therefore it's impossible to connect the neutrinos coming from A and B to make them virtual and achieve neutrinoless double beta decay.

be conserved, in beta decay an antineutrino must be emitted, so that the arrows flow continuously from the neutrino to the electron (recall Figure 19.4). Normal double beta decay is just two such disconnected diagrams happening at the same time (see Figure 26.3). If the neutrino is Dirac, there's no way of connecting the two diagrams, say, by linking the neutrinos (forming a virtual line between the two decays,) because their arrows would clash. A emits an antineutrino and wants it to be absorbed by B. But this would mean B emitting a neutrino, whereas instead it emits an antineutrino (see Figure 26.3). Neutrinoless double beta decay is therefore not possible in Dirac's picture.

The story is altogether different with Majorana neutrinos, explaining Wendel Furry's insight. In Feynman diagrams Ettore's neutrinos are represented idiosyncratically because they flow both ways in time, representing particle and antiparticle

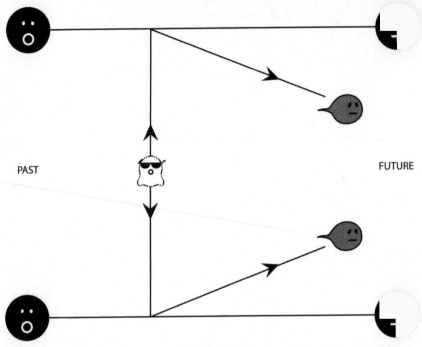

PAST   FUTURE

Figure 26.4: But Majorana neutrinos are "bi-timal": Their time flow, or their particle-antiparticle control, goes both ways down their lines. Thus the two beta-decay diagrams can be linked via a virtual neutrino without a clash. The neutrinoless double beta decay is the hallmark of the Majorana neutrino.

in one. Thus they're represented as lines with arrows pointing both ways (recall Figure 19.5.)

We can now see why neutrinoless double beta decay is possible should the neutrino be Majorana. Because the time flow, or the particle-antiparticle control, goes both ways down their lines, we can link the neutrino lines belonging to the two diagrams without a clash (see Figure 26.4). The Majorana neutrino can annihilate itself, and effectively that's what happens inside the diagram of neutrinoless double beta decay. It's the sign of the beast: of Ettore's beast. And that's how one hopes to bring it to light.

So here's a blast from the past. Even as we delve into twenty-first-century neutrino experiments, we return to where it all started: the spectrum of beta decay. Physicists work by analogy, so it's interesting to find out what was on their minds during those days long ago when Pauli and Ettore were first messing about with neutrinos. When I started my physics studies, it was firmly believed that the neutrino was massless; but at its inception, in the 1930s, it was never meant to be massless. It was quickly realized that its mass had to be much smaller than the electron mass. Still, no one assumed it to be zero. But how did they know that the neutrino mass had to be so small?

Recall my early diagram, showing how the energy "spectrum" of beta rays pointed Pauli toward a third man, the neutrino: the carrier of vanished energy. You had two nuclei with different masses, the daughter being lighter than the mother by ΔM. This means that an equivalent energy E (equal to ΔMc^2) was available, and should by rights have been carried by the electron; i.e., the beta ray. The expectation was for the beta ray spectrum to be a sharp line around this energy, which was the same for each mother and daughter pair (see Figure 26.5). Instead the beta ray spectrum showed a range of possible energies, all smaller than this amount, following a smooth line (see Figure 26.6).

Energy conservation was salvaged by Pauli's proposal of the neutrino. The observed electron spectrum revealed that the electron was sharing the available energy with a mystery particle—invisible, intangible. Near point A of the spectrum, the neutrino carried all the available energy, E, so there was nothing left for the electron. Near point B, the electron instead carried the lot, the neutrino being left with nothing. In between, they'd share energy E. Right?

Almost. Point B, the "end-point of beta decay," as we call it nowadays, represents the minimum energy that can be given to the neutrino (it is the maximum energy we ever see the electron carrying). This is only the full E if the neutrino is massless, because then you can give it an energy as close to zero as wanted. But if the neutrino is massive you have to give it at least its rest mass. Therefore the maximum you can give the beta ray is not E, but E minus the neutrino rest energy ($E_\nu = m_\nu c^2$); see Figure 26.7.

Sounds easy? Let me put the picture to scale (see Figure 26.8). To this day no one has measured the neutrino mass using this method. We do know, however, that it has to be much smaller than the electron mass, indeed at least a million times smaller, because the end-point of beta decay is *very* close to point B.

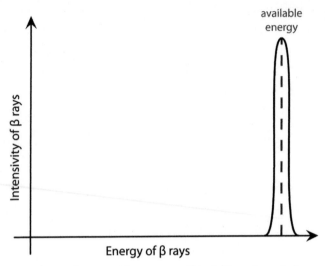

Figure 26.5: The expected spectrum of (single) beta decay, before physicists were forced to face up to the neutrino.

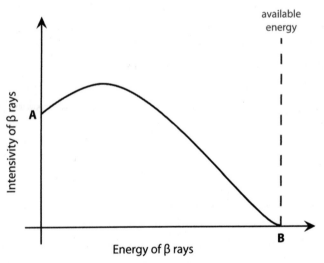

Figure 26.6: The spectrum of (single) beta decay as it was observed. At point A, the neutrino carries all the available energy (so that none is left for the electron, or beta ray.) At point B, the neutrino carries none.

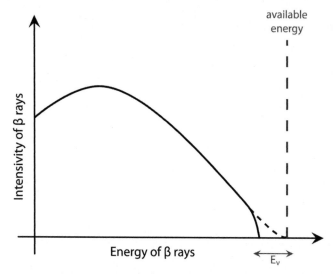

Figure 26.7: A correction for what happens at B, given the neutrino rest mass (and so its rest energy $E_\nu = m_\nu c^2$.) The neutrino must carry at least this energy, so the beta ray can never absorb all the available energy. The end point of beta decay, in principle, would offer us a method to measure the neutrino mass.

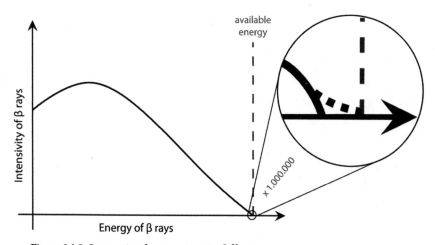

Figure 26.8: In practice the story is quite different.

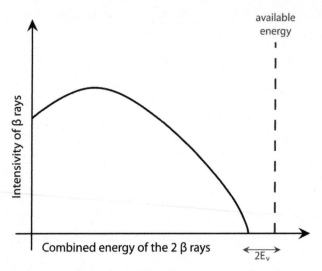

Figure 26.9: The spectrum of double beta decay with production of neutrinos. When the energies of the two electrons (beta rays) are added up we observe a curve starting at zero and ending at the available energy minus twice the neutrino mass (since we must now feed two neutrinos).

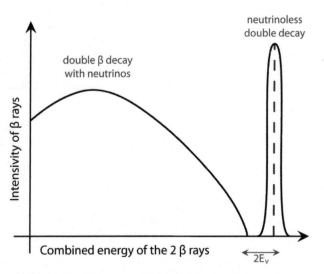

Figure 26.10: The spectrum of neutrinoless double beta decay, instead, reproduces the expectation for the spectrum of beta decay before the neutrino entered the picture: a sharp line centered at the available energy.

But in this context, the spectrum of *double* beta decay is precisely a combination of the energy spectrum scientists expected to see in 1930 and that which was in fact observed. Life is a long circle. If double beta decay occurs with production of neutrinos, when we add up the energies of the two electrons coincidentally emitted we get a smooth spectrum, starting at zero and ending at the available mass-energy ΔM minus twice the neutrino mass (see Figure 26.9). But if no neutrinos are emitted we get, of course, what was expected before Pauli was forced to dream up the neutrino: a sharp line at the available energy (see Figure 26.10). Ironically, the signature of the Majorana neutrino is precisely what physicists expected before they knew about neutrinos—because it marks beta decay without them. Majorana's neutrino leaves its mark by *not* being emitted, because it self-annihilates, thus failing to imprint on the spectrum the energy deficit which revealed the neutrino in the first place.

It's bizarre that the sign of the beast is self-annihilation. Ettore's neutrino is the most self-destructive particle in the cosmos—like its creator.

TWENTY-SEVEN

Ettore Majorana™

In 1966 Edoardo Amaldi wrote *The Life and Work of Ettore Majorana*. In 1974 director Leandro Castellani shot a film on Ettore and published *Dossier Majorana*. In 1975 Leonardo Sciascia wrote his classic *La Scomparsa di Majorana*. Ettore's letters and documents were published by Erasmo Recami in 1987. Since then, Italian books on Ettore have sprouted by the dozen. If you add the hundreds of newspaper articles putting forward conspiracy theories, plus TV documentaries, comics, and novels, Ettore can hardly be said to lack an afterlife. The search for neutrinoless double beta decay can only envy such a level of attention.

Many have tried writing about Ettore in fiction format. It has always been disastrous: In a fictional world the temptation to invent answers is just too great. In trying to speculate beyond the facts, Ettore always becomes a Mickey Mouse character. In *L'inglesina in Soffitta* he inexpertly has sex with a Scottish spy girl (in Italian novels British women are invariably portrayed as the ultimate tarts, which is, of course, unfair.) In a French book he runs away to Argentina where he becomes a family man; one of his daughters is named after Maria.

Then there are the "real-life" stories purporting to solve Ettore's mystery. A prime example: Signorina Tebalducci, a Florence lady who claimed in the Italian press to have dated him. According to her, it was not a happy affair even by the standards of the day (when vaginal penetration had to be preceded by a blessing at the altar). Ettore didn't speak much, and whenever things became interesting in the slightest, a group of "foreign-looking men" would turn up, take him aside and talk for hours, completely ignoring her. After discussing the matter with a brother in the Carabinieri, the signorina concluded that she was being used as a *"schermo,"* a statement that underlies the theories portraying Ettore as a spy, kidnapped by the enemy in 1938.*

The mathematically oriented clochard who once lived in Mazara del Vallo.

Ettore's brother Luciano vigorously dismissed this story, stressing that Ettore had never been to Florence. But Signorina Tebalducci performed a remarkable public utility service. She filled in the gap that other accounts of Ettore leave wide open: his private life. A lie is certainly less infuriating than a void.

Personally, I like to collect Majorana conspiracies the way others collect stamps: There are nearly as many. My favorite concerns *"l'omu cani,"* Sicilian for "man-dog." In 1940 a drifter appeared in Mazara del Vallo, in southern Sicily, not far from the setting of the detective Montalbano stories. The self-named Tommaso Lípari claimed to come from Africa and carried with him a profusion of bags. He never begged and didn't accept money, slept in a cave at a nearby ruined castle, and kept his distance from everyone. He chain-smoked, crafting his cigarettes from butts he picked up from the ground. And he could do cubic roots in his head.

One morning in 1973 the man-dog appeared dead. "Natural causes" was entered in the official report. To everyone's surprise his funeral attracted "notorious VIPs" (excuse the euphemism). It further transpired that Signore Lípari had a deep scar on his right hand (not necessarily associated with crashing a vehicle) and wore a bracelet with Ettore's birthdate engraved on it. But what set imaginations alight was that the case attracted the attention of anti-Mafia hero Paolo Borsellino.

---

* I particularly like her use of the term *schermo*. It's a beautiful word, with many senses. It means "a screen" or "a shield" but it's also used by Dante in *Beatrice* to signify his feelings of modesty, hiding the sacred terror that Monna Bicce—the nine-year-old who had conquered his heart—inspired in him. Majorana's "schermo" is also both the root of his attraction and a cover for our unmitigated shame.

Borsellino was born and raised in the Palermo district of La Kalsa, where they tell tourists not to go, and had been a childhood friend of that other anti-Mafia hero Giovanne Falcone. Aware of their surroundings, they put themselves through law school, so that one day they could cripple the Mafia. The results are well-known: They now have an airport named after them. Both died from car bombs. The blast hitting Falcone opened a ten-meter-deep crater, and the highway had to be redesigned. You can't run away from death, but you can be prepared for it. Falcone and his wife decided not to have children because they didn't want to leave orphans.

Borsellino concluded in his report that Signore Lipari was not Ettore Majorana. Unwittingly he cast the Mafia aspersion on both.

Whoever the man-dog was, people missed him when he died. In Mazara del Vallo he's still invoked by adults to frighten misbehaving children. If someone goes mad people say that "he's become worse than the man-dog." And if you need a cubic root evaluated, things have never been the same.

This is only one of the theories in my collection, and to be honest most strike me as disingenuous: nothing like a good conspiracy to attract attention. There are exceptions, of course; sometimes they're just so alluring. I can't forget the conversations I had with my friend Professor Francesco Guerra who, having spent hours scorning Ettore theories, couldn't resist telling me his own.

He found a letter from Florence physicist Gilberto Bernardini to Ettore's friend Giovanni Gentile Jr. The letter is not dated, but practical details included place it around May 1938 (it refers to a trip Gentile would take to Germany in June 1938). It's a reply to an earlier letter from Gentile (presumably reporting on the events of March) and opens with "As you can imagine, the news about Majorana has filled me with great joy. It's not a pretty thing, perhaps, but it's not as tragic as we had imagined. We can all be happy about it."

If this letter was indeed penned in May 1938, how can we interpret it? No two ways about it. *Obviously* Ettore didn't commit suicide (not as tragic as they had imagined). He did something not very "pretty" (ran away from his family and responsibilities), but it could have been significantly worse (i.e., suicide). They're all well aware of his fate, and can be "happy about it," even filled with joy.

Unfortunately it's not so obvious. The dating is soft. The letter could also have been written at the end of 1937 (the practical details referred to can be accounted for in other ways). In which case the letter simply refers to the mundane shenanigans of the *concorso* (not very pretty to suspend it, but they all got accommodated in the end). But one can't suppress a glistening in the eye. This letter embodies the lingering sensation that friends and family know what happened but aren't telling. My sincere congratulations to them.

But these somewhat measured conspiracy theories pale in comparison to the all-out flights of lunacy. That Ettore's family is cursed; that aliens have abducted him; that he ran away to the center of the Earth via Etna, where he's working for the benefit of humankind. Shall we ratchet it up a notch in abject insanity? If so, let's indulge in an Internet myth I've added to my collection: the "curse of the Majoranas."

During recent restoration works in the Steri (the ancient Palermo headquarters of the Inquisition) a variety of graffiti emerged from beneath four centuries of plaster, left by inmates due to be tortured and killed. Some of them were signed by Paolo Majorana, who was there in 1681. Understandably, his graffiti contain nothing but gritty curses.

In a certain town in the middle of Amazonia political and economic life is controlled by the Majoranas. The family patriarch has been nicknamed Majorana, O Mafioso, and many people who've been aggrieved by his prepotencies have admitted that they're too scared to take their cases to the police. Instead they've resorted to good old-style Brazilian voodoo.

On November 12, 2003, a group of Italian Carabinieri were killed by a suicide bomber in Nassiriya, Iraq.* Among the fallen heroes was Orazio Majorana. The premature death of Orazio affected the family enormously, but no one more so than his brother, as chance would have it, named Ettore. Ettore Majorana became a vocal pro-peace activist, denouncing the hypocrisy of Western intervention, and demanding the recall of all Italian forces. But after a flurry of political activity, young Ettore buckled to depression, never quite getting over his bereavement. A few months later, the twenty-two-year-old modern Ettore Majorana appeared drowned. Close friends believed it was suicide, but no one really knows. This creates a bizarre parallel between the two Ettores: one left apparent suicide notes but no corpse; the other a corpse but no suicide notes.

---

* An incident that petrified a nation under the belief that her soldiers were sent abroad for the purpose of parading medals and singing opera.

PIZZE DA € 4.00

Turi Ferro	mozz.di bufala,radicchio,grana,speck, olio,pomodoro
Angelo Musco	mozz.tonno,cipollina,pomodoro fresco peperoni,origano olio
Vincenzo Bellini	mozzarella,salame picc.,cipollina pomod.fresco peperoni,origano,olio
Giovanni Verga	mozzarella,spinaci,pancetta stufata,panna grana,olio
Nino Martoglio	pesto,mozzarella di bufala,pomodorini origano,olio
Ettore Majorana	mozzarella di bufala,funghi freschi,crudo rucola,grana,pomodoro
Pippo Baudo	pepato fresco,speck,cipollina,olive origano,mozzarella

Pizza Ettore Majorana: Buffalo mozzarella, fresh mushrooms, Parma ham, rocket, "grana" cheese, and tomato.

The fact that none of these three Majoranas is related to our Ettore hasn't stopped abundant Internet literature from uniting them under the "curse of the nuclear physicist," the "curse of the Majoranas," and their evident relation to the "curse" of a certain combustible baby. Perhaps in this context it's not so odd that a scientific article appeared on the WEB archives stating that Ettore had placed himself in a quantum superposition of dead and alive, like Schrödinger's cat. And that another article associated his name with a neutrino device capable of deactivating all the nuclear weapons on Earth.

But then, by now Ettore has become a trademark. He has a pizza named after him. He's a hero in comics, surrounded by cats in the fifth dimension. People manage

What is it, boys? What are you mewling at?   There's no one there. . . .

to keep a straight face while arguing over who owns the copyright to his suicide notes. Ettore has fallen prey to those who insist they have exclusive rights to his soul, those who allege they own the patent for oral sex.

He doesn't need neutrinoless double beta decay for anything. In the vast seas of the subculture, his afterlife is in full swing.

TWENTY-EIGHT

A Vote of Silence

✣ ✣ ✣ ✣ ✣ ✣ ✣ ✣

But we're close to the end: Shall we step back down a few notches in lunacy? Something akin to an answer is expected, after all. So let me tell you the Majorana fable closest to my heart. A Vote of Silence. I'm an atheist, but still I find it irresistibly beautiful.

One evening in the 1960s, the novelist Leonardo Sciascia was having dinner at a Roman restaurant with Emilio Segrè and writer Alberto Moravia, among others. The conversation touched upon the atomic bomb and Segrè, in good Basilisk style, boasted of his participation in the Manhattan Project. It was at the height of the cold war, with the phantom of global destruction hanging over the world, and such lightheartedness was even more pathetic than irresponsible. Sciascia, who had the hot blood of all Sicilians, shouted verbal abuse and attempted to physically assault Segrè. Moravia had to restrain him lest Emilio end up with his spectacles crushed into his nose. And it was during this mealtime incident that Sciascia—an antinuclear activist—first conceived the notion that scientists are evil. From direct contact with a Dr. Strangelove.

As Sciascia read about the politics and science behind the making of the atom bomb, he developed his views further. As he learned more about those who made the bomb and their dilemmas of the soul—or lack thereof—he became convinced that scientists like Fermi, Segrè, and Oppenheimer were actually clinically mad. But then he stumbled upon Ettore. Immediately Sciascia realized that he was different. Ettore had a soul. His life wasn't just science without any padding of humanity. Sciascia began collecting documents on Ettore's life obsessively. He speculated that Ettore being a Sicilian—a nation that copiously produced artists and writers but no scientists—may have had something to do with it. Sicilians are superstitious and despise science. A Sicilian scientist *had* to be different from the others.

Some time later, Sciascia was reading a newspaper while relaxing at a café. His mind was fluttering away when he came across an article reporting that one of the pilots who dropped the bomb had subsequently joined a monastery. And that's when he had his sudden vision, his burst of insight! Of course that's what happened to Ettore. In 1938 he'd seen what the world was coming to, and unlike the others he couldn't shake off responsibility. He cherished his freedom and didn't want to be a mere robot programmed to inaugurate the nuclear nightmare. So he'd sought refuge at a monastery.

Unlike other Ettorologists, Sciascia never claimed to have holy fire spewing from his ass; indeed, he rejected the label of "theory," deeming his views on Ettore "a mixture of history and invention." Ettore, in his mind, became symbolic of the scientist who refuses to give up choice, who repudiates being a cog in a political mechanism.

"I have no doubt that Majorana would have participated in the Manhattan Project." This is Professor Francesco Guerra's opinion.

"Why?"

"Out of curiosity. He'd have wanted to know if the physics of a chain reaction really was possible." I suppose this is why scientists look so deranged to the rest of the world; and the man in front of me is the nicest person one can think of.

I have no doubts that Ettore would never have participated in the Manhattan Project. Even before moral issues are brought into the picture, he could never be part of a group. *Any* group. It just wasn't in his nature.

Maria Majorana explained long ago how the family repeatedly petitioned the Pope for information on Ettore, always receiving no reply. Certainly if Ettore did join a monastery (as suggested by several clues gathered by the 1938 search party), that must have left a paper trail, which may eventually resurface from that black hole called the Vatican. And Sciascia himself found a scent of Ettore at a monastery in Basilicata.

The first glass of wine I ever drank was in the grounds of a Carthusian monastery near Evora, Portugal. My uncle had a friend who worked at the attached farm, and we were invited for lunch. Befittingly for a vineyard that produces some of the best nectar in the world, the only drinks available were tap water and wine. My uncle asked that I be given water (I was seven or eight) but when the meal arrived this offended the farmers' aesthetical sense so much that they insisted on supplying me with a robust glass of wine.

None of this veneration of the sublunar world is permitted on the other side of the wall, in the monastery itself. A vow of silence is observed. One of my earliest memories is the sound of their bells tolling at three or four in the morning calling the monks to their prayers. The monks are divided into brothers and fathers (or choir monks); the brothers are allowed meager contact with the outside, but the fathers are totally detached from it. Choir monks spend their days in spiritual exercises: praying, meditating, and studying.

It was in one of the monasteries of this order that Sciascia found the evidence he didn't need for what he was happy to see as a metaphor. "Perhaps someone in this place was able to avoid betraying life by betraying the conspiracy against life," he writes, describing his visit.

He finds the peaceful and tidy cemetery, row after row of past monks orderly arranged in their final launch pad to Heaven; but no inklings as to Ettore's location. A Dutch monk shows him around, insisting that the order has no famous writers or scientists to boast of. This contradicts the story that led Sciascia there, a friend who visited in 1945 and vividly recalled being told that there was among the choir monks a famous scientist.

But as Sciascia follows the monk through the austere citadel he admits:

"I have no wish to ask questions, to verify. I feel implicated, compelled to respect a secret. . . . On the threshold, as I take my leave, the Carthusian asks:

"'Have I answered all your questions?'

" . . . I asked but a few. . . . But I answer:

"'Yes.'

"And it's true."

Could Ettore have guessed what was about to befall the world? Almost certainly yes, and this merely adds another stratum to the crisis that hit him from 1933 to 1937 and finally unwound itself in those stormy days in March 1938. Ettore had a tradition of spotting what the other Panisperna Boys couldn't and they did achieve the fission of Uranium in 1934, having inflicted cancer on the poor goldfish of their pond. Ida Novak, in Germany, understood it in 1934, so why not Ettore? But unlike Novak (who was a chemist), Ettore would also have been aware of the prospect of a chain reaction and the formidable energies that could be released.

Whether he foresaw Armageddon is another matter. But he was so depressed in those years—possibly, or even probably, for other reasons—that any human extrapolation he'd make from such a discovery would have to have been pessimistic. I don't believe that the bomb was *the* cause of Ettore's depression, but I do think that it was one among many culprits. Enough to turn the wildest epicure into a monk, if you want my opinion.

The ending of the film *I Ragazzi di Via Panisperna* re-creates Sciascia's vision almost to the letter. Recall that in the film Ettore withdraws from the world, becoming a peasant at a bucolic Sicilian farm south of nowhere. Fermi visits him ostensibly to beg his assistance. He's carrying a notebook, "Uranium" embossed on its cover.

When they meet the scenario is surreal. We see Ettore ambling through the dilapidated farmhouse, hefty chandeliers hanging from the branches of a palm tree that grew embedded in a neoclassical reception room. Birds fly freely indoors. Ettore receives Fermi warmly, but replies to his requests for help with "I don't occupy myself with that anymore." His statement echoes an utterance from Arthur Rimbaud, the teenager who changed poetry forever but never wrote another verse after he turned twenty.

Ettore then mentions Galileo's classic "Il Mondo Mobili," which proposed that the Earth orbits the sun and led to his troubles with the Inquisition. "I like Galileo because he was afraid," says Ettore (what could he have been thinking when he said this?). Fermi responds with the banality that the truths discovered by Galileo are universal facts and would always have been found, if not by Galileo, then by someone else. Ettore sniggers:

"We only find what we want to find."

A bewildered Fermi leaves his uranium notebook next to a pile of loquats and they part forever.

That night, Ettore can't sleep. The scene is almost mystical, flooded by Sicilian a cappella songs, with their Arabic inflexions and extraterrestrial rhythms. Amid nightmarish flashbacks of his mother forcing her Ettoronzo to perform circus tricks before visitors, he sweats and is delirious, then gets up flushed and distraught. With a violent sweeping gesture he makes space on a messy desktop and opens Fermi's notebook. Effortlessly, he solves the enigma of uranium, and then—predictably—he burns his work.

After Ettore vanishes in March 1938, Fermi travels to Naples and learns from Carrelli that two days before evaporating, Ettore had drawn five months of salary from the university cashier. Among Ettore's belongings, spread out by Carrelli on a table, Fermi recognizes "Il Mondo Mobile." Lost in its pages is a picture of Ettore. At the gates of a monastery.

From the picture Fermi locates the monastery, where he meets a monk well acquainted with Ettore. But Ettore, the monk says, has left.

"Have I arrived too late?"

"We can't measure the time of God with our laws."

The monk gives off a clear whiff that he knows Ettore's whereabouts but won't talk. Before Fermi takes his leave the monk gives him a book. It's his uranium notebook, which Ettore had asked the monk to return to its rightful owner, should he ever visit.

Soon afterward, Fermi moves to the United States with his family. As they cross the Atlantic, one night Fermi is so upset he's insomniac. He goes out to deck to watch the sea, and Laura joins him. Fermi invokes Ettore's memory, praising his gifts. Laura observes:

"Ettore had exceptional qualities. But in other ways he was handicapped. He lacked the most basic common sense. He couldn't have ended up in any other way."

"So you too believe that he committed suicide?"

"I want to doubt, but can't."

And then Fermi's final words, closing the film, drop out:

"Ettore wanted to disappear. But even in that, he was a genius. He wanted to disappear leaving us the uncertainty of not knowing whether he's dead or alive. So that, whatever happened to him, effectively he *is* alive. We'll always feel as if he's looking over our shoulders, judging us in everything we do. Imposing on us the impossibility of ignoring our conscience."

A shame that, in reality, Ettore's shadow didn't prevent Fermi from helping to build the bomb. As Pirandello so characteristically put it, "The dead are the pensioners of remembrance."

EPILOGUE

Mediterranean Whales

After Ettore's disappearance, all the characters in this book had distinguished lives, even if not always for the best reasons. If Ettore didn't commit suicide, he must have enjoyed reading about them in the papers.

Ettore's friend Werner Heisenberg has sometimes been accused of knowing more about Ettore's fate than he ever let on. Some even claim that Ettore rejoined Werner in 1938 to give him a helping hand. Because something remarkable happened to Heisenberg six years after Ettore visited Leipzig. He was rehabilitated by the Nazi regime.

In 1933 Heisenberg had opposed Deutsche Physik, leading to a major schism in the regime. God knows what pressures he was under when Ettore visited. He was inevitably against equating relativity with "Jewish physics" and other such imbecilities. But at the same time he was a right-wing ultranationalist and must have been pained at being rejected by such a pro-German movement as the Nazis. By comparison his collaborator Pascual Jordan was enjoying life as a Nazi storm trooper. Ambivalence and a certain jealousy must have been Heisenberg's feelings toward the regime in the period between 1933 and 1938. Regrettably, his bond to his country and its people was indeed tantamount to a bond with the Nazis.*

So in 1939, when the Nazi establishment finally ruled in his favor and against Deutsche Physik, he jumped with open arms at the offer of leading the German atomic project—the so-called Uranium Club.

---

* All material and quotes in this epilogue regarding the atom bomb are taken from Richard Rhodes's excellent *The Making of the Atom Bomb* (see References at the end).

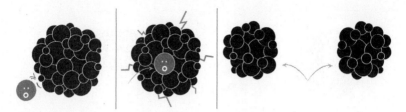

Figure E.1: The mysterious process that takes place when uranium is bombarded with neutrons: The neutron is absorbed and a much distressed nucleus splits into two roughly equal-size pieces.

Later Heisenberg justified himself by saying that under the prevailing winds he was certain Hitler would win the war and Nazism become an unavoidable nuisance. He only wanted to safeguard the values of science and culture under the order of the "Thousand Year Reich." Poor excuse. Particularly as this was not about Greek philosophy and the solipsism of quantum mechanics—he could have been the *reason* Hitler won the war. Asked by Hitler's advisers how big the new bomb being built by his club would be, he replied, "The size of a pineapple" (strange choice of fruit). If Signorina Tebalducci got it right, Ettore was much involved in the fruit salad.

Uranium, I don't need to tell you, went on to become a major character in this play. Ida Noddack was right when she argued in 1934 that bombarding uranium with neutrons need not produce heavier elements. Instead it causes a dramatic fission into two much smaller pieces (krypton with $Z = 36$ and barium with $Z = 56$), releasing a large amount of energy (see Figure E.1). The process was only unambiguously revealed in 1938 and 1939 by Lise Meitner, Otto Hahn, and Fritz Strassmann. Lise Meitner was an Austrian Jew, in a serious predicament after the Anschluss, forcing her to do a runner to Sweden while this important discovery was being made. Thus both sides of the conflict knew about the new phenomenon at first hand.

Yet, uranium had a panoply of further twists in store, and we shouldn't criticize Fermi too hastily for having missed them. The Boys, of course, had been wrong, but in 1934 nothing heavier than an alpha particle had ever been chipped from a nucleus. Moreover, this made complete sense using the nuclear theory of Ettore and Heisenberg, which everyone revered at the time. A full revision of nuclear theory was required, and I have to wonder how aware of this Ettore might have been, so critical was he of his own work.

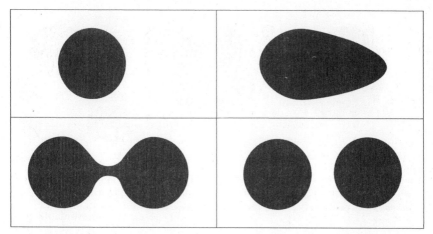

Figure E.2: A revision of nuclear theory was required in order to accommodate the possibility of fission: the liquid drop model.

In the end, the new theory was provided by Meitner's nephew Otto Frisch and the not-so-senile Niels Bohr. Surprisingly, uranium behaves like a "liquid drop" when fission takes places (see Figure E.2). But even with this extraordinary revelation, the mystery wasn't fully solved. Fission was confirmed independently by several groups but the full fission dataset was self-contradictory, particularly concerning the dissimilar actions of fast and slow neutrons. Even the identity of the two fragments wasn't fully clear (they eventually turned out to be krypton-92; i.e., with A = 92, and barium-141). Uranium was a real mess.

Physicists puzzled and scratched their heads . . . until Niels Bohr—again!— had an epiphany (obviously Carlsberg is good for you). Playing a game of analogies with other elements and the behavior of liquid drops, Bohr remembered that uranium exists in nature as a mixture of two isotopes: about 99.3 percent has A = 238 (Uranium-238) but 0.7 percent has A = 235 (uranium-235). He then understood in a flash that when uranium is bombarded with slow neutrons—such as those first exposed at the Via Panisperna pond—*it is the minority component that undergoes fission.*

This means that three neutrons had to be emitted; just count the neutron and proton content of mother and daughter nuclei (see Figure E.3). To Bohr's shock, he realized that these neutrons were then free to restart the process, hitting other uranium nuclei and initiating a chain reaction: 1 fission, 3 fissions, 9, 27, 81, 243, 729, etc. . . . (see Figure E.4). All that was needed was to purify the minority isotope

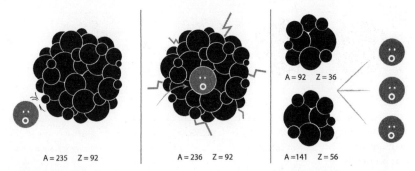

Figure E.3: The minority component, uranium-235 (92 protons and 143 neutrons) hit by a neutron splits into krypton-92 (36 protons and 56 neutrons) and barium-141 (56 protons and 85 neutrons.) This leaves three spare neutrons, which are emitted.

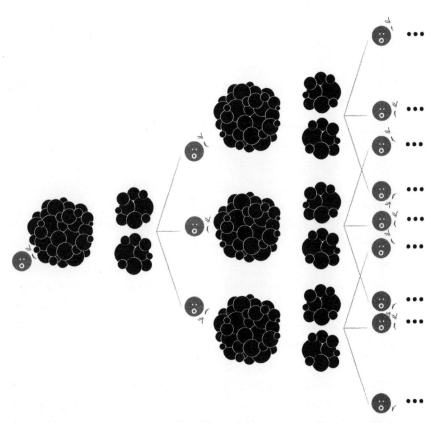

Figure E.4: The leftover neutrons released by each fission are now free to attack new uranium-235 nuclei to induce further fissions and more free neutrons, sparking off a chain reaction.